Communities of
Learned Experience

Singleton Center Books in Premodern Europe

The Charles S. Singleton Center for the Study of Premodern Europe of the Johns Hopkins University is an interdisciplinary consortium of humanities scholars on the university's faculty. Established in 2008, the Singleton Center fosters research of the European world in the Late Classical, Medieval, Renaissance, and Early Modern periods. The Singleton Center sponsors graduate research abroad and faculty-led initiatives to partner with European institutions of higher learning, as well as educational activities, lectures and lecture series by prominent scholars, and many other scholarly activities on the Johns Hopkins campus in Baltimore and at European venues. The center is named after Charles S. Singleton (1909–1985), the renowned scholar of medieval literature who taught for most of his career at the Johns Hopkins University.

Every two years the Singleton Center organizes the Singleton Distinguished Lecture Series, which invites a prominent scholar of premodern Europe to the Homewood campus of Johns Hopkins to present three lectures on a common theme. The series was inaugurated in October 2010 by Professor Nancy Siraisi on the subject of epistolary networks in Renaissance medicine. The present volume is based upon those lectures.

LAWRENCE M. PRINCIPE
Director, Singleton Center

COMMUNITIES *of* LEARNED EXPERIENCE

Epistolary Medicine in the Renaissance

NANCY G. SIRAISI

The Johns Hopkins University Press
Baltimore

This book was brought to publication with the generous assistance of the
Singleton Center for the Study of Premodern Europe.

2 4 6 8 9 7 5 3 1

The Johns Hopkins University Press
2715 North Charles Street
Baltimore, Maryland 21218-4363
www.press.jhu.edu

Library of Congress Cataloging-in-Publication Data

Siraisi, Nancy G.
Communities of learned experience : epistolary medicine in the
Renaissance / Nancy G. Siraisi.
p. ; cm. — (Singleton Center books in premodern Europe)
Includes bibliographical references and index.
ISBN 978-1-4214-0749-4 (hbk. : alk. paper) — ISBN 978-1-4214-0784-5
(electronic) — ISBN 1-4214-0749-3 (hbk. : alk. paper) —
ISBN 1-4214-0784-1 (electronic)
I. Charles S. Singleton Center for the Study of Premodern Europe.
II. Title.   III. Series: Singleton Center books in premodern Europe.
[DNLM: 1. Lange, Johannes, 1485–1565.   2. Augenio, Orazio,
ca. 1527–1603.   3. Correspondence as Topic—Europe.   4. Physicians—
history—Europe.   5. History, 16th Century—Europe. WZ 140 GA1]
610.92'24—dc23   2012008903

A catalog record for this book is available from the British Library.

*Special discounts are available for bulk purchases of this book. For more
information, please contact Special Sales at 410-516-6936 or
specialsales@press.jhu.edu.*

For Desmond and Niko Siraisi

CONTENTS

Acknowledgments   *ix*

Introduction   1

1 Contexts and Communication   14

2 The Court Physician Johann Lange
and His *Epistolae Medicinales*   38

3 The Medical Networks of Orazio Augenio   62

Conclusion   85

Notes   *89*
Index   *155*

I was extremely honored by the invitation to give the first set of Singleton Lectures at the Charles Singleton Center for the Study of Pre-Modern Europe of the Johns Hopkins University in October 2010. This book presents a revised version of those lectures. At Johns Hopkins I should like to express my gratitude especially to Christopher Celenza, codirector of the Singleton Center; to Lawrence Principe, director of the Singleton Center; and to Gianna Pomata of the Johns Hopkins Institute of the History of Medicine. I am also grateful to Lawrence Principe for guidance regarding arrangements for publication. My thanks to Walter Stephens for helpful discussions regarding a section of chapter 2. Gianna Pomata and an anonymous referee for the Johns Hopkins University Press offered insightful and helpful comments on the manuscript of this work, and I am also indebted to Gianna Pomata for suggesting the title adopted for the book.

I am most grateful to Anthony Grafton, for providing helpful comments on an earlier draft of this work, and to David B. Ruderman, who read a draft of chapter 2 and contributed valuable suggestions. And, as always, I have learned much from the work of Ian Maclean, who has pioneered the study of published letter collections as a genre of early modern medical books. I thank Arlene Shaner, acting curator of the Rare Book Room of the New York Academy of Medicine, for her unfailing helpfulness. Remaining mistakes and omissions are, of course, my own.

At the Johns Hopkins University Press, I am especially grateful to Jacqueline Wehmueller for her thoughtful and gracious editing.

COMMUNITIES OF
LEARNED EXPERIENCE

Introduction

THE NUMEROUS Latin letters on medical subjects written by sixteenth-century physicians constitute one small part of a much larger universe of early modern European learned and scientific correspondence. For a generation now, the role of correspondence in early modern erudition and exchange of ideas, of arguments, of objects of interest to collectors has been a focus of much historical and historiographical attention. Numerous and valuable studies have been directed to such topics as humanistic epistolarity, the concept and realities of the Republic of Letters, and the correspondence network of a celebrated individual.[1] Recently, the sixteenth-century rise and spread of the practice of publishing printed collections of medical letters, explicitly designated in the title of the volume as *epistolae medicinales* or something similar, has also begun to play a part in this historiography. Ian Maclean has provided a valuable overview and analysis of such collections by twenty-one different authors published between 1521 (the date of the first edition of the *Epistolae medicinales* of Giovanni Manardo, the humanist physician of Ferrara whose work initiated the genre of printed collections of medical letters) and 1626, with attention not only to content and authors but also to printing history.[2] Among recent studies of the correspondence of individuals, the work of Candice Delisle on the letters of the physician and naturalist Conrad Gesner takes thoughtful account of their dissemination (whether as individual items transmitted beyond the original recipient or collected in a printed volume) and of the process of editing them for publication.[3]

Detailed studies of the correspondence of individual physicians (such as in the last-named example) can throw much new light not only on an author's medical interests and activities but also on his local situation and loyalties (political or religious), intellectual and scientific milieu, and network of correspondents. Latinate physicians in sixteenth-century Europe, particularly those who had received their medical formation in universities, can, in general and broadly speaking, be considered as belonging to a scientific community. Most members of that community had many aspects of medical theory and practice in common, yet they might gain their life experience of medical activity in a variety of social and professional settings—for example, those of the university, of the court, and of urban medical practice—and inhabit very different geographic, religious, and political environments. In short, collections of medical correspondence produced in these various milieux illuminate the writers' different worlds of learned experience as well as their ideas and activities.

Within the compass of the present work it is possible to indicate only a few examples of what such collections have to offer historians—and not just historians of medicine and science but also cultural, intellectual, and (to some extent) social historians. But since it is evident that most printed collections of Latin medical letters published in the sixteenth century came either from Italy or from the German lands (though many of the Italian collections were subsequently republished in northern Europe), it seems appropriate to examine some instances from each geographical region and, in particular, to look at any examples of medical correspondence exchanged between regions. Especially in the latter such cases, the presence or absence of references to religious issues or to disputed medical innovations, most notably Paracelsianism, is likely to be of interest to others besides historians of medicine strictly defined. In the age of Reformation and Counter-Reformation, the negotiation of religious boundaries affected every area of intellectual life; at the same time, in the realm of natural philosophy and medicine, the three principles, ideas about separation, and alchemical therapies of Paracelsus and his followers represented a radical departure from Aristotelian-Galenic learned tradition. Accordingly, following this introduction, chapter 1 presents some examples of and considerations relating to medical letters sent between regions of Italy and the German lands. The two succeeding chapters discuss, respectively, letter collections by one German and one Italian author, each of whom kept his correspondence almost entirely within a regional network, broadly defined.

The number of editions of collected Latin medical letters published over the course of the sixteenth and early seventeenth centuries is in itself sufficient to justify their classification as a recognizable minor genre of Renaissance learned medical discourse. Not only did numerous authors contribute to the genre, but the collected medical letters of some writers appeared in multiple and successively expanded editions. The many-times-reprinted *Epistolae medicinales* of Giovanni Manardo (1462–1536) of Ferrara provides a salient, but not unique, example: over the course of twenty years this collection grew from six internal books contained in a slender quarto in the first edition of 1521 to the twenty internal books in the folio edition of 1540.[4] As a genre, medical letter collections share the attention to experience increasingly present in other types of contemporary medical writing, as well as the flexibility and variety characteristic of Renaissance miscellanies in various fields. In addition, they embody, at least to some extent, the notion of free communication among equals inherent in the concept of the Republic of Letters. But it is also true that the genre is a loose one, encompassing considerable diversity among individual collections in type of content, style, and organization. Indeed, some authors stretched the definition of *epistola* in their printed collections of *epistolae medicinales* to include not only letters to named addressees but also essays with no addressee, humanistic dialogues, and treatises divided into multiple chapters.

Printed collections of medical letters thus merit attention as a variety of Renaissance medical literature, even though they are far from providing the full picture of contemporary correspondence networks among physicians and naturalists (most of whom in the sixteenth century were trained in medicine). Some correspondence networks—involving exceptionally distinguished individuals, famous among their contemporaries—were both larger and more geographically widespread than any printed medical letter collection could reflect: the naturalist and bibliographer Conrad Gesner (1516–1565) exchanged letters with a least 240 individuals in several different parts of Europe; many hundreds of letters from more than a hundred correspondents addressed to the encyclopedist Theodore Zwinger (1533–1588) survive in the library of Basel University; the Clusius Project at Leiden University has now digitized the more than 1,500 surviving letters that the botanist Carolus Clusius (1526–1609) received from some three hundred individuals and then carefully preserved. Such correspondence linked a widespread professional community, but its members could hardly have been familiar in any but the most formal sense. All three of these men were trained in medicine, and Gesner and Zwinger were professors and practitioners of medicine; each of their correspondence networks included many physicians; but of the three, only Gesner was the author of a collection of *epistolae medicinales* printed in the early modern period (assembled and edited after his death).[5]

Recent studies note the extent to which early modern physicians and naturalists contributed to, and valued, newly available descriptive information of many kinds, whether about natural objects or about products and practices originating in remote regions; moreover, they point to a new emphasis on the actual or potential utility of such information as well as its overall contribution to knowledge.[6] From another standpoint, participation in correspondence networks offered prized opportunities for attitudes and activities that have been described as characteristic of the Republic of Letters: reciprocal friendship and open exchange (of specimens, of information about the world of nature and of learning, of honorable mention in each other's works), even if the ideal of harmony was at times interrupted by sharp disputes when rules of proper behavior were perceived to have been broken.[7] For example, the first letter in Gesner's published *Epistolae medicinales*, addressed in August 1561 to the imperial physician Johann Crato of Krafftheim, fulfills all the requirements for friendly sharing of information: it thanks Crato for his letters, specifying when each one was received; informs him of the movements of the printer Pietro Perna and the forthcoming publication of the medical *consilia* (advice on therapy for individual patients) of Gian Battista Da Monte, the recently deceased professor at Padua; lets him know that the Montpellier professor of medicine and naturalist Guillaume Rondelet has stopped writing letters (at any rate to Gesner) now that he is preoccupied with his young wife and with editing his (Rondelet's) own works; gives an acerbic account of Pietro Andrea Mattioli's progress with his labors on Dioscorides and briefly justifies Gesner's disagreements with Mattioli over aconite and other matters; and tells of Gesner's recent visit to thermal baths for his health. The letter also encloses and promises gifts: with it comes a list of some German works of Paracelsus, accompanied by condemnatory remarks about occult sciences and Paracelsus's presumed association with demons; while a copy of a work of Gesner's own, currently in press, in which Crato is mentioned in the preface, is promised.[8] Gesner's letter to Crato thus exemplifies forms of scholarly communication central to the concept of the Republic of Letters, as well as recalling contemporary teaching about "the art of writing letters," as Erasmus entitled his celebrated manual on the subject.[9]

Another correspondent of Gesner opined that the sharing of medical information by letter served the purposes of the Christian as well as the learned republic. Luigi Mondella, under the persona of a speaker in one of his *Medical Dialogues* (1551), reported that Gesner—"a man erudite in every field and very friendly to me"—had recently sent him some writings on alchemy (*chymica ars*) from the library of a recently deceased physician; the other interlocutor in the dialogue is made to respond that the information would be "for the benefit in the first place of the *respublica Christiana* and then for all nations."[10] The claim that Gesner transmitted alchemical material has verisimilitude, for Gesner—despite his hostility to Paracelsus—was keenly interested in alchemy, considered as a practical art for preparing medicaments by distillation.[11] Mondella, a physician from Brescia, was himself the author of a collection of his own medical letters, published in several editions.[12] Two of his letters to Gesner, probably from 1552, survive in manuscript at Zurich; Gesner, in turn, included in his ornithology (1555) appreciative acknowledgments to Mondella both for providing information about quail found in one of the latter's published collected letters and for sending a picture that helped Gesner identify another avian species.[13]

Like the medical dialogue quoted above, Mondella's medical letters contain a notable element of religiosity. In them it took the form not only of frequent invocations of Christ but also of a vigorously expressed objection on religious grounds to the practice of prescribing different medicines for rich and poor, to which Mondella added a strain of social as well as religious criticism with the remark that Christ himself, as the son of a carpenter, would not have been eligible for any position of civic dignity in sixteenth-century Brescia. Perhaps this element in Mondella's writing reflects the heightened religious atmosphere in Brescia at a time when the city, then under Venetian rule, was simultaneously affected by reformed religious tendencies from northern Europe and by the countermeasures of the Catholic Reformation.[14] Probably nothing about Mondella's own religious position should be read into his correspondence with the Zwinglian Gesner and preference for publishers in Zurich and Basel; the terms of warm friendship in which he wrote in 1542 to the Carmelite friar Angelo Castiglioni, who in the 1550s was to become noted for his activities against heresy, would seem to indicate Catholic allegiance—yet even this evidence is ambiguous, as Castiglioni himself was described as having had Nicodemite (i.e., crypto-Protestant) tendencies in the 1540s.[15]

A different emphasis emerges from the remarks of a compiler of medical letters writing toward the end of the century. Lorenz Scholz von Rosenau, a physician and botanist from Silesia, assembled two vast compilations of, respectively, medical *consilia* and medical letters by many different contemporary or recent physicians. Within a few days during March 1598, he wrote prefaces for both collections.[16] In his preface to the collection of *consilia*, he stressed the practical usefulness for medicine of preserving written *consilia* as a record of cases of disease and their treatments.[17] And in the preface to the collection of medical letters titled *Philosophical, Medical, and Chymical Letters of the Most Distinguished Philosophers and Physicians of Our Age*, he emphasized the comparable value of medical letters as a vehicle for disputation or debate on medical issues. Scholz explained and justified the enterprise of letter collecting as follows:

> In my judgment nothing shows the usefulness of letters in medicine to be any less than that of *consilia*. Letter writing is an ancient method of inquiring about a proposed problem, disputing about it, and of responding to someone asking and inquiring, and also of examining whatever the response consists of. Rhetoricians in the schools of dialectic call such letters dialectical, questioning, and didactic. And it is this kind of letter, familiar indeed but not vulgar, that is especially proper to learned men. For if they strongly doubt something or do not agree among themselves in their opinions, they are accustomed to dispute by means of letters on various subjects sent here and there.

As distinguished recent examples of such letters, Scholz called attention to one by Pico della Mirandola and one attributed to Melanchthon, writings that were on opposite sides of the debate over the respective merits of philosophy and rhetoric. With a thrust at humanist writers of letters on philology in ancient literary texts, Scholz added: "How many letters on disputed places and variant readings in [ancient] authors have not recent critics, of whom our century is almost too prolific, published? If this is appropriate about things of minimal, indeed trivial importance, indeed insignificant words, what will we think ought to be done by us in medicine if something occurs which produces doubt in us?" Scholz went on to enumerate writers of serious, meritorious letters in antiquity: Greek and Latin theologians, Cicero, Plato, and Hippocrates. He concluded with a list of sixteenth-century authors who had written letters of significance for the discipline of medicine and whose collected letters had appeared in print: eight Italian physicians who were authors of collections titled *epistolae medicinales* or something similar, together with Gesner, Crato von Kraff-theim, the Heidelberg professor of medicine and theologian Thomas Erastus, and the Heidelberg court physician and medical humanist Johann Lange.[18]

With these words Scholz neatly placed his collection in all the relevant contexts. The reference to the letters universally believed in the Renaissance to have been written by Hippocrates attested the antiquity and venerability of the medical letter as a genre.[19] The allusion to Melanchthon, in conjunction with a list of medical authors who (for the most part) were Italian, calls to mind the Republic of Letters as it supposedly extended across political, religious, and disciplinary boundaries in its search for truth. And Scholz's insistence that the usefulness of the letter was equal to, but different from, that of the *consilium* enrolled physicians' letters among relatively recent developments internal to medicine. *Consilia*, or written advice for individual patients (often addressed to the patient's medical attendant rather than directly to the patient), began to appear in the thirteenth century and multiplied by the fifteenth. Although *consilia* certainly represented a somewhat empirical aspect of medieval and Renaissance European medicine, their content usually consisted primarily of recommendations for therapy; they seldom described the course of disease and even less often reported its outcome (the separate early modern development of the medical case history, tracing the course of disease from onset to outcome, whether favorable or not, owed much to the Renaissance interest in *historia* in all its senses and to the recovery of the complete Hippocratic *Epidemics*).[20]

Of course, some medieval and Renaissance physicians wrote letters, as well as *consilia*, long before the sixteenth century, although, as one might expect, for the most part only correspondence involving exceptionally distinguished patients appears to have survived.[21] But the publication of large assemblages of medical letters on diverse themes seems to have been a sixteenth-century innovation. Letter collections, like other agglomerations of varied items treating subjects in short compass, appealed to the taste for miscellanies that humanistically educated physicians shared with their humanist contemporaries. And, like the *Observationes* studied by Gianna Pomata, if to a lesser degree, much in the letter collections revealed and responded to a growing professional and scientific interest in shared accounts of personal experience and in *autopsia* (seeing for oneself).[22] To the extent that letter collections came to include debates over unresolved or contested issues in medicine, perhaps in some instances they, like humanist dialogues, embodied a certain openness to the perception that truth might emerge through discussion from more than one viewpoint (though medical letters also include many examples of vitriolic personal controversy).[23] Nor was Scholz was the first to compare the medical *consilium* with the letter. As Ian Maclean has noted, Manardo, in the defense of the epistolary form with which he opened his letter collection, had already drawn a similar parallel: "This way of writing through letters . . . is not new. According to Galen, Archigenes wrote eleven books of medical letters, and according Paul [of Aegina], Themison wrote ten books. Moreover, what the *recentiores* call *consilia* are certainly nothing other than letters."[24] But whereas Manardo had simply insisted that *consilia*, too, were in effect letters, Scholz emphasized the different role of those letters that—unlike *consilia* for patients—were vehicles for learned debate and disputation among physicians about medical and natural philosophical theory and practice. That, in actuality, published collections of *epistolae medicinales* often also included many *consilia*, and that physicians evidently continued to find collected *consilia* useful, did not invalidate this distinction.[25]

At the same time, Scholz's reference to the definitions of types of letters by "rhetoricians in the schools of dialectic" directly linked medical letters to the humanist attempt to revive ancient epistolary forms. Humanist enthusiasm for the recovery of antiquity extended to ancient letter collections, most notably those of Cicero and Pliny, and in turn stimulated a number of humanists, beginning most famously with Petrarch, to form compilations of their own letters. Over the course of the fifteenth and early sixteenth centuries, instruction in letter writing became part of rhetorical education, and numerous authors—of whom the most celebrated were Erasmus and Vives—produced manuals on the subject. In general the humanist manuals sought to replace the medieval *ars dictaminis*, which taught elaborate formulae of address mostly used in official letters of various kinds, with a more informal, personal style. Familiar letters, the cliché went, should be brief and should resemble a conversation between absent friends.[26] No doubt physicians, like most other members of the literate and Latinate classes, encountered such instruction in the early stages of their humanistic education. But as Scholz noted—perhaps recalling Erasmus's brief mention of an epistolary "disputatoriae genus" or the various kinds of substantive letters recognized by Vives and Lipsius—instructors in letter writing also acknowledged a category of didactic letters intended to teach or to support serious arguments.[27] In this last category, in Scholz's view, letters addressing debates among physicians about serious issues in medicine were at least as worthy of publication as were collections of essays titled *variae lectiones* (a term with a much broader application in the sixteenth century than its literal modern equivalent, "variant readings"), or something similar, about trivial textual problems in ancient authors.[28]

But despite (or perhaps because of) his careful keying of his collection to central and well-established themes relating to humanist letter writing, medical letters, and the Republic of Letters, Scholz's project was unusual in one important respect. His comments occur at the beginning of a large anthology that brought together the medical letters of many different contemporary physicians from several parts of Europe. The model for Scholz's project was perhaps one of the few earlier examples of an anthology of medical letters by multiple authors, namely the collection *Epistolae medicinales diversorum authorum* of 1556, since the writers listed in the full title of that work correspond, in the same order, to the first five names on Scholz's list of authors of medical letter collections.[29] By contrast, most sixteenth-century printed collections of medical letters announce themselves as the work of a single individual. Authors often prepared (and doubtless edited) their own collections of letters for the press, as did the botanist and Habsburg court physician Pietro Andrea Mattioli (1501–1578) and the Paduan professor of medicine Orazio Augenio (ca. 1527–1603). The latter made clear his commitment to readying his own letters for publication in the dedication of a third volume of them: "These were the reasons why I will now publish very little of the labors I described above, but in their place this third volume of my *Epistolae medicinales*."[30] In other instances a family member or a close associate undertook the task of editing, presumably as a celebratory or commemorative enterprise; such was the case for the Paduan professor Vettor Trincavella (1496?–1568), whose letters were compiled by his son, and for Conrad Gesner, whose *Epistolae medicinales* were redacted posthumously by Kaspar Wolf (also the editor of many of Gesner's other works), whose labors included writing to Gesner's correspondents to request the return of letters.[31] In an earlier project Scholz himself had edited multiple volumes of the *Consiliorum et epistolarum medicinalium* of the imperial physician Crato von Krafftheim—although in Scholz's hands the collections named for Crato also had the character of an anthology, as they included many letters to and from members of Crato's circle as well as those by Crato himself.[32]

Sixteenth-century authors of Latin letters on medical topics and controversies wrote from a variety of professional settings, not all of which were academic. Throughout the century the most celebrated centers of medical education were in Italy, and it was in Italy that both medical humanism and the vogue for printed collections of medical letters had first spread. But from midcentury on, in northern Europe as well as in Italy, authors of published letter collections included court physicians and town physicians as well as university professors of medicine—indeed, the trajectory of a learned medical career might at different times encompass all these occupational settings. (Hence it is difficult if not impossible to assign fixed characteristics to sixteenth-century medicine in these different contexts: a growing empiricism emerged in all of them; and innovators were found in academic circles as well as among urban or court practitioners, as illustrated by the careers of a number of the most prominent figures in the history of sixteenth-century medicine.)

Like other categories of early modern humanistic or scientific correspondence, manuscript letters on topics of professional interest often circulated beyond the original addressee. No doubt some correspondence among physicians remained a private exchange, neither intended nor sought out for wider circulation in any form. But letters containing observations about striking cases or a discussion of current topics of medical debate, especially if by an author of renown, were probably the most likely to be passed on. It was presumably through such informal transmission of individual items that Scholz, clearly an especially enthusiastic collector, was able to assemble the large corpus of material making up *Philosophical, Medical, and Chymical Letters of the Most Distinguished Philosophers and Physicians of Our Age.* Yet the transmission of some manuscript medical letters among members of the learned medical community does not necessarily indicate either that the physician authors of the letters in question deliberately chose manuscript over print for the circulation of some of their works or that the learned medical profession in general functioned as a familiar coterie (as the term has been applied to some early modern literary authors and circles).[33] Rather than a coterie, Latinate physicians formed a professional community in the sense that they had many features of their academic training in common. But they could be, and many were, at some remove from one another—divided by their degree of openness to critiques of Galen and other standard medical authorities; by their type of medical activity; and by social, religious, and regional contexts. Moreover, it is abundantly evident, and deserves to be emphasized, that physicians who assembled and edited collections of medical letters—whether their own or, like Scholz, those of others—sought out publication in print and viewed dissemination in print among a broader professional learned community as an advantage.

Thus, to note only a few examples, Philippe Tingus declared in his dedicatory preface to the 1556 *Epistolae medicinales diversorum authorum* that nothing was more useful to *studiosi* of medical matters than medical letters "that discuss the darkest shadows of the art with wonderful dexterity"; he went on to recall that the dedicatee had lamented that these had to be read scattered in "as many books as there are authors" and suggested that it would be so much better if such letters were gathered into one volume; as a result, he (Tingus) had arranged for the present anthology of letters by various authors to be published by the Lyon branch of the house of Junta. The author of the preface to a posthumous reprint of the letters of Johann Lange noted, "You [the editor of the volume] wanted to see that distinguished work of medical letters . . . again brought to light, and indeed to receive that new light from the house of Wechel the outstanding publisher of our

Germany: which for elegance of typeface and diligence in work, and singular pleasure given by its publications, deserves praise and commendation." Theophilus Mader, a pupil of the physician and theologian Thomas Erastus, in the preface to his edition of the latter's *Medical Disputations and Letters*, both provided specific examples of the medical usefulness of the letters and recorded the help he had received from Erastus's widow in assembling them for publication in print.[34]

Printed collections of medical letters had the potential to reach a wide professional audience, though their distribution was, as Ian Maclean has pointed out, partly determined by the preferences of certain publishers for this type of material.[35] Many of the collections attributed to single authors contain only one side of a correspondence; in a number of such cases (though not in that of Gesner) their authors or editors presumably made use of copies of letters retained by the sender, perhaps for the purpose of forming just such a compilation. But there are exceptions, an important one being Mattioli's *Epistolae medicinales*, which includes letters both to and from him.[36] And, as will become apparent, the letter collections edited by Scholz also contain some examples of both sides of an exchange. But even when a volume is limited to just the sender's correspondence, the names and positions of addressees offer much information about the professional, intellectual, regional, and confessional networks—and sometimes more intimate and familiar coteries—in which the writers of medical letters were embedded. For these writers, as much as for any of the famous humanist letter writers from Petrarch to Erasmus, their collected letters served to provide a document of the author's own understanding of his life and times.[37]

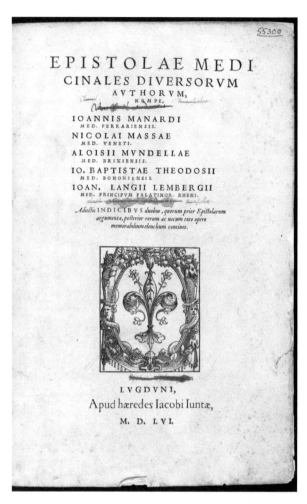

EPISTOLAE MEDI
CINALES DIVERSORVM
AVTHORVM,
NEMPE,

IOANNIS MANARDI
MED. FERRARIENSIS.

NICOLAI MASSAE
MED. VENETI.

ALOISII MVNDELLAE
MED. BRIXIENSIS.

IO. BAPTISTAE THEODOSII
MED. BONONIENSIS.

IOAN. LANGII LEMBERGII
MED. PRINCIPVM PALATINOR. RHENI.

Adiectis INDICIBVS *duobus, quorum prior Epistolarum argumenta, posterior rerum ac uocum toto opere memorabilium elenchum continet.*

LVGDVNI,
Apud hæredes Iacobi Iuntæ,
M. D. LVI.

Medical Letters by Various Authors (Lyon, 1556), Wellcome Library, London. Each of the authors included had previously published a collection of his letters as an independent item. The issuance of these collections, assembled into one volume, may be an indication of the publisher's awareness of growing interest in this type of medical writing. Courtesy of the Wellcome Library.

CHAPTER ONE

Contexts and Communication

THIS CHAPTER CONSIDERS some examples of letters exchanged in the middle and later decades of the sixteenth century between learned physicians in Italy and in northern Europe that subsequently came to be included in contemporary printed collections of medical letters. Though its primary concern is with letters as they appeared in printed collections, something first needs to be said of the material and political conditions likely to have affected the transmission—and perhaps to some extent and, in some cases, the content—of the original correspondence on which the printed collections drew. In the second half of the sixteenth century, letters between northern parts of Italy and the German lands could be exchanged by various means. Overland travel between these regions was frequent, primarily because of established patterns of long-distance trade; in addition to merchants, other travelers included northern students drawn to the Italian universities and, in some famous instances, Italian religious exiles fleeing north to areas of Reformed religion. Some courier services existed, or missives might be entrusted to merchants or other travelers.

But such correspondence was not without hazards. Letters and those carrying them had to travel across the Alpine passes and through different linguistic regions, the territories of different rulers, and often—in the age of Reformation and Counter-Reformation—regions of different religious adherences. And written communications could and did go astray through robbery or could fall into the wrong hands. Moreover, those in Venetian territories who wrote to and received letters from correspondents abroad were doubtless aware of the intermittent efforts at censorship of Venetian political authorities, at times in conjunction with Catholic religious authorities.[1] Nevertheless, the common intellectual formation and sense of professional identity that bound together academically trained Latinate physicians in general (as well as the European renown and reach of the University of Padua) provided the background for a good deal of medical correspondence across the Alps and across political and, at times, religious boundaries.

Several of the Italian participants in these exchanges were members of the faculty of medicine of the University of Padua, then famous throughout Europe as a center of medical learning and education. Another was a physician and author who practiced and taught medicine in Venice, a city where such private teaching provided many opportunities for students to gain experience in clinical practice. Their northern correspondents included both learned physicians associated with universities—professors at Heidelberg and Zurich and a student at Basel—and physicians at the imperial court, first in Vienna and later in Prague.

The motives that might encourage any individual physicians to engage in medical correspondence with colleagues in distant regions were no doubt as numerous and diverse as the physicians themselves. Yet some characteristics common to many of these communications emerge. The Republic of Letters was above all an intellectual community, and undoubtedly many physicians used letters to their counterparts in other parts of Europe primarily as a vehicle for engaging in spirited—and often heated—long-distance debate about disputed issues in medicine or related areas of natural philosophy. But it clearly appears that such correspondence was also, at least for some physicians some of the time, a practical, career-oriented exercise—that is, part of an endeavor to attain hoped-for professional benefit or advantage. Knowledge of new books published in other regions, of innovations in theory or practice, and of the movements or appointments of leading physicians in courts and universities elsewhere was useful information in multiple ways. At the same time, as letters circulated, whether from hand to hand as individual manuscript items or in the form of a printed collection by a single author published during his lifetime, they could serve to advertise the author's scholarship and medical expertise as well as the range of his interests and contacts. Moreover, for members of university faculties of medicine, epistolary contacts with colleagues in distant centers of learning could be helpful as a way of recommending—or keeping track of—students, as attested by some of the letters between Theodore Zwinger in Switzerland and Girolamo Mercuriale in Padua (discussed later in this chapter). And, of course, some long-distance medical correspondence had a basis in personal ties or family obligations. In an age in which many northern European medical students traveled south to attend Italian universities, and when some young medical graduates undertook a *peregrinatio medica*—a tour of important centers of medical learning, seeking out distinguished physicians—direct acquaintance among medical students or physicians from widely separated regions of Europe was by no means uncommon.[2] Such contacts were, in the nature of things, likely to be short lived, but they might well facilitate or inspire subsequent correspondence.

A Venetian Practitioner and His Family

Examples that fall into several of the categories just mentioned come from the *Epistolae medicinales et philosophicae* of the Venetian physician Nicolò Massa. Massa was a prosperous and energetic medical practitioner and teacher in Venice whose publications on topics of current medical interest included treatises on anatomy, *morbus gallicus* (syphilis), and pestilential fevers. He also published two volumes of "medicinal and philosophical letters": the first, containing thirty-four letters, appeared in 1550; the collection was reissued with the addition of a second internal volume containing twenty-seven more letters in 1558. Massa's correspondence ranges over a variety of topics and two items in it are notable for their acerbic comments on current developments in anatomy—but as Richard Palmer has pointed out, a substantial part of each volume directly or indirectly concerns either Apollonio or Lorenzo Massa, two nephews for whose care, upbringing, and education the elder Massa became responsible. Apollonio, a central figure in the first volume, became a physician like his uncle. Lorenzo, the recipient of a number of letters in the second volume, had his uncle's support and assistance in entering the service of the Venetian state.[3] Yet the letters Nicolò Massa wrote in 1552 and 1553 to Lorenzo, at that time an arts student at Padua, are detailed expositions of small sections of the *Canon* of Avicenna, then still standard as a university medical text—topics that suggest that Nicolò thought it important for this nephew, too, to be medically well informed, even though Lorenzo did not develop a professional vocation in this field.[4]

But it was Apollonio's studies that provided Nicolò Massa's contacts with the world of German medicine. Apollonio Massa graduated in medicine from Leipzig, where reformed religion had recently been introduced and where he studied with a well-known Lutheran professor of medicine, Heinrich Stromer of Auerbach. Moreover, Apollonio also matriculated at Wittenberg, although his uncle's published letters do not say so.[5] Sixteen of the thirty-four items in Nicolò Massa's first volume of *Epistolae medicinales* are essays on medical topics and are addressed either to his nephew, to other Italian physicians, or, in a couple of cases, to Venetian notables; another fifteen are *consilia* for patients (two of them with German names); and one other "letter" is in reality a brief treatise on dialectic. (By comparison, in the second volume of Massa's letters the proportion of *consilia* increased to seventeen of twenty-nine items. Seven of the seventeen are for cases of *morbus gallicus*, a preponderance that presumably reflects Massa's reputation for expertise in this area as well as the success of his treatise on the subject, printed five times in Latin and twice in Italian translation during his lifetime.)[6]

But the remaining two letters in the first volume, both of them to Auerbach (addressed respectfully by Nicolò Massa as "senex doctissime"), are very different in character from the rest: their topics are, respectively, the creation of the world and the immortality of the soul. Massa apparently wrote both letters in response to Auerbach's queries about his opinions in these matters.[7] In the first, Massa firmly proclaimed his belief in God's creation of the world, emphasized the limits of human knowledge, and refuted counterarguments about the eternity of the world. In the second, he similarly both asserted his own belief in the immortality of the soul and claimed that Aristotle could be interpreted in the same sense, while noting that some illustrious Aristotelian philosophers believed the soul to be mortal and that some of the moderns, naming Pomponazzi among them, had given their own interpretations of Aristotle's views.

So far as I am able to judge, in these letters Nicolò Massa himself espoused a very traditional and entirely orthodox Catholic, Christianized Aristotelianism. Moreover, his nephew Apollonio, who perhaps found Leipzig and Wittenberg attractive in the first place because of an interest in reformed religion and who certainly received his medical education in

Lutheran circles, presumably also remained Catholic. At any rate, Apollonio subsequently enjoyed a successful medical career in Venice without, so far as is known, encountering any religious censure.[8] But since Auerbach's Lutheran commitment could hardly have been unknown to Nicolò Massa, his interchange of views on such topics with this addressee remains striking. In all probability Massa's willingness to correspond with Auerbach on philosophically and religiously sensitive issues can, at one level, simply be related to the ease and the many modes of contact—intellectual, economic, diplomatic, and personal—between inhabitants of the Venetian Republic and those of Europe north of the Alps. Venice's proximity to routes carrying travelers, trade, and correspondence to and from northern Europe and its status as a major Mediterranean port ensured that, especially in the first half of the century, it remained a mart of ideas and information, as well as of goods, from many sources. Moreover, at the nearby University of Padua, the most important academic institution in Venetian territory, students of northern European origin, who organized themselves as the German Nation (*Natio germanica artistarum*), constituted a significant part of the student body.

But Massa's mention of Pomponazzi in connection with the subject of the immortality of the soul clearly situates this letter in the context of contemporary philosophical controversy and religious concerns. At the University of Padua, a major center of Aristotelian philosophy, the topic of the immortality of the soul had been very prominent earlier in the century. Both Pietro Pomponazzi, the author of a famous treatise denying that the soul's immortality could be proved by reason, and his chief opponent, Agostino Nifo, had been educated and had taught at Padua before moving elsewhere. When Pomponazzi published his treatise on immortality in 1516, it generated both strong opposition from Nifo and papal condemnation.[9]

In a more general context, efforts to enforce religious conformity intensified by midcentury, in reaction to developments in northern Europe. A number of prominent individuals from Italy and from southern cantons of Switzerland fled north to territory more hospitable to Reform during the 1540s and 1550s; they included Bernardino Ochino and Pietro Martire Vermigli, both well-known members of Catholic religious orders with long careers ahead of them as theologians and controversialists in various Reform circles.[10] These religious exiles included another writer of medical letters, the physician Taddeo Duno (formerly a student of Girolamo Cardano at the University of Pavia), who, in a famous episode, led a band of religious exiles from Locarno over the mountains to Zurich.[11] And in Venetian territory confessional lines were being firmly drawn. In 1547, five years after the establishment of the Roman Inquisition, the Venetian government instituted its own magistracy for the repression of heresy. By the time Massa's letter collection was published in 1550, discussion in Venice of theological issues—or of natural philosophical topics possibly touching on theology—was probably already more circumspect than it had been in the early 1540s, when his letters to Auerbach were written.[12]

Medical Debates in the 1570s and 1580s: Members of the Medical Faculty at Padua and Their Northern Correspondents

A more expansive and wide-ranging type of interregional medical correspondence is exemplified by letters exchanged between some Paduan professors and leading physicians in northern Europe in the 1570s and '80s. Unlike Massa's correspondence with Auerbach, almost none of these letters were published in Italy; when they found their way into print it was in letter collections published in the German lands. The letters between German and Paduan physicians in these collections are different in character from those that Massa had included a generation earlier in his *Epistolae medicinales et philosophicae*. Massa's published letters to correspondents in Germany were all either *consilia* for individual patients or somehow connected with his personal concern for his nephew. By contrast, the published correspondence by other authors seems to have been predicated on academic connections and reputations and is mainly concerned with general issues and theoretical questions in medicine, rather than with individual patients or cases. Lorenz Scholz assembled numerous such letters on general topics in the collection that he named for Crato von Krafftheim (though in fact it also contained letters by many other authors) and published in multiple volumes beginning in 1591; he republished many of the same items in his more broadly titled anthology of medical letters published in 1598.[13] Professors of medicine at Padua exchanged letters both with the imperial court physician Johann Crato and with Crato's disciple and associate Peter Monau (or Monavius). Similar letters to and from some of the same Italian professors of medicine are also to be found among the posthumously published letters of Thomas Erastus, professor of medicine at the reformed University of Heidelberg, remembered both as a Zwinglian theologian and, in medicine, as a vigorous opponent of Paracelsianism.[14] One theme running through a number of these letters, written in the 1570s and '80s when both Italy (notably Venice in 1575) and other parts of Europe were suffering from a series of epidemics described as plague, was controversy over the transmission of disease, and specifically over the concept of contagion.[15] Though contagion was certainly an important and a recurrent theme, it was by no means the only topic of discussion in these exchanges.

From 1574 to 1576 Erastus corresponded with two prominent professors at Padua: Bernardino Paterno, holder of a senior chair in medical theory from 1563 until his death in 1592, and Girolamo Capodivacca, who began to teach at Padua in 1553, held a senior chair in *medicina practica* from 1561 until his death in 1589, and earned a reputation as a friend of the organization of northern students (the German Nation).[16] The correspondence between Erastus and the professors at Padua was initiated from Padua. On September 20, 1574, Erastus received, via a letter from Theodore Simmelbecker (a former Heidelberg student now at Padua), a copy of Capodivacca's objections to Erastus's "theses on contagion," together with Paterno's "very learned cogitations" on the same. It is not clear what work containing Erastus's views had come into the Paduans' hands. Possibly Capodivacca and Paterno had received a manuscript copy of the "Theses de contagio" (dated 1574), "respondente M. Timotheo Madero Helvetio," which first appeared in print many years later in Erastus's collected *Disputationum et epistolarum medicinalium volumen*. Timotheus Mader (1541–after 1592), along with his brother Theophilus Mader (another of Erastus's students and later the editor of his *Disputationes et epistolae*), received the master of arts degree at the University of Heidelberg in 1568. Or were the Paduans perhaps responding to Erastus's *De causa morborum continente tractatus*, published along with one of his anti-Paracelsian works by the exiled Italian printer Pietro Perna in Basel in 1572?[17]

Simmelbecker, the former Heidelberg student, explained in his cover letter to Erastus that Capodivacca "wants to discuss the subject [of contagion] with you by letter, since he cannot do so in any other way, and asks that you will write back as soon as possible." Erastus complied with great promptness. His approximately 2,500-word defense of his opinions to Capodivacca is dated two days later, on September 22.[18] Another of about the same length to Paterno followed two days after that. Capodivacca had argued that, properly speaking, no disease could be called contagious, since a disordered *complexio* (the balance of primary qualities—hot, wet, cold, and dry—in the body), assumed by Galenists to be the explanation of internal illness, was not in itself contagious. Erastus, too, maintained the role of complexional imbalance in all internal disease but insisted in addition that one category—pestilential disease—generated noxious and contagious vapors (51v). In the same letter Erastus considered whether and in what circumstances *spiritus* (in the ancient medical sense of a refined material substance in the body) could putrefy. The same themes occupied his response to Paterno, as well as two additional, lengthy letters to Capodivacca and Paterno, again written within a few days of each other, in February of the following year.[19]

In another long letter to Capodivacca in September 1575, Erastus again discussed the role of putrefaction in disease and repudiated the concept of "disease of the total substance" put forward by two medical writers with a contemporary reputation as innovators, the professor of medicine at Paris and royal physician Jean Fernel (1497–1558) and Giovanni Argenterio (1513–1572), who taught at the University of Pisa and elsewhere.[20] Fernel was the author of a famous treatise *On the Hidden Causes of Things* and of a comprehensive textbook of physiology, pathology, and therapy that remained one of the most widely read general works on medicine until well into the seventeenth century. He positioned himself as a reformer of medicine, although for the most part he espoused a modified Galenism. Fernel nevertheless played a significant role in undermining Galenic disease theory through his denial that imbalance in the body's temperament, understood as the mixture (*complexio*) of the four primary qualities, could account for all diseases. His alternative explanation, which drew on his ideas about hidden forces in nature to postulate a separate category of "diseases of the total substance," was one component in the long series of sixteenth-century controversies about the causes, nature, and transmission of disease.[21] By contending that some diseases, especially pestilential epidemics, were caused by a specific quality that affected the body as a whole, Fernel and his followers were able to insist on the distinct identity of such diseases. Giovanni Argenterio similarly acquired a reputation as a sharp critic of Galen's disease theory in particular, and of excessive reliance on ancient authority in general.[22]

Erastus's correspondence with Capodivacca was not undertaken solely for the sake of highly theoretical disquisitions on contagion and disease theory, rife with analogies about the transmission of substance, or quality, from one thing to another. It also had a practical side, as is apparent from a letter of January 1576 in which Erastus thanked Capodivacca for kindness to the German students whom he, Erastus, had recommended for study at Padua, especially Peter Monau.[23]

Erastus's letters indicate that Capodivacca replied several times, but only one of the replies, dated 1576, is found among the letters edited by Scholz. It reiterates Capodivacca's position that no disease could be said to be contagious *formaliter*, since Galen and Avicenna both asserted that the cause of disease was complexional imbalance. The same letter implies that Capodivacca was beginning to regret having initiated the correspondence: "Because you are trying with a long oration, and wonderful skill, and great diligence to refute my opinion about contagion and the putrefaction of *spiritus*, and the other things that I gladly wrote to you about, I will now *briefly* [my italics] state my position[.]"[24] The correspondence between Capodivacca and Erastus seems to have ceased at this point, but Capodivacca and Monau, whom Erastus had recommended to Capodivacca as a student, continued to correspond for several years after Monau left Padua to finish his medical studies at Basel. These letters, too, contain much theoretical discussion about causes of disease, but—unlike anything in Erastus's letters to Capodivacca—they are also quite strongly oriented toward medical practice. In addition to repeatedly asking his former teacher for fuller explications of some of his general recommendations about therapy, Monau sought Capodivacca's advice about the nature and proper treatment of a current epidemic.[25]

Girolamo Mercuriale as a Writer of Medical Letters

But by far the most celebrated member of the medical faculty at Padua to correspond regularly and frequently with colleagues in northern Europe in the 1570s and 1580s was Girolamo Mercuriale. Both Capodivacca and Paterno, despite their senior chairs at a major university, were medical authors of fairly modest renown. Capodivacca produced a general treatise on medical practice, reprinted several times, and also acquired a certain reputation for his writing on and treatment of *morbus gallicus*. Paterno's principal work was a commentary on the first section of the first book of that Arabo-Latin medieval standby of medical education, the *Canon* of Avicenna. Although based on lectures on a text still assigned in the Padua medical curriculum, Paterno's commentary on Avicenna was nevertheless a decidedly old-fashioned contribution to the literature of medicine by the time of its publication in 1596.[26] By contrast, Mercuriale—at Padua from 1569 until 1587, where he was Capodivacca's colleague as professor of practical medicine—was a medical scholar and prolific author of much greater distinction, as well as a renowned practitioner whose patients at different times included Cardinal Alessandro Farnese, the Emperor Maximilian II, and Grand Duke Ferdinand of Tuscany; his fame among his contemporaries was such that the Duke of Urbino requested a portrait of the distinguished physician. Among Mercuriale's many publications were a major antiquarian work on ancient exercise, *De arte gymnastica libri sex*; a whole series of treatises based on his lectures on specific medical topics—diseases of women, diseases of children, diseases of the skin, poisons, and plague; and an edition of the Hippocratic corpus.[27]

At Padua, Mercuriale's responsibilities included the official capacity of "protector" of the *Natio germanica artistarum*—that is, the association of students of arts and medicine from many parts of northern Europe who came to the University of Padua. Relations between the *Natio* and Mercuriale deteriorated sharply in 1579, when, jointly with the bishop of Padua, he issued a set of rules obliging the northerners (many of whom were Protestants) to respect Catholic observances. The northern students responded by denouncing Mercuriale and requesting protection from the Venetian government.[28] But outside Padua, Mercuriale's reputation for goodwill toward northerners endured. During the time Gaspard Bauhin spent at

Padua as Mercuriale's student, he had made a copy of a manuscript of lectures on diseases of women given by Mercuriale some years earlier. In 1586, after returning to Basel, Bauhin edited a collection of gynecological works, among which he included Mercuriale's lectures. In his preface to Mercuriale's treatise, Bauhin asserted optimistically that even though the edition was produced without Mercuriale's knowledge ("eo inscito editus sit"), the latter's friendly attitude to "Germans"—that is to say, members of the *Natio germanica* at Padua—ensured his approval.[29] In reality, as Michele Colombo (another former student) explained, Mercuriale "at first indeed was not very pleased, because earlier he had denied the opportunity to publish the material to Venetian, Paduan, and foreign publishers who had been demanding it importunately. But afterward, since he was kindhearted and very well disposed to Germans, he took it in good part." Nevertheless, Mercuriale authorized Colombo to prepare an edition of the same lectures, published the following year in Venice, with a title making clear that it was both approved by Mercuriale and intended to supersede Bauhin's version.[30]

In addition, Mercuriale was a prolific writer of letters: his range of learned correspondents extended beyond medicine to the Florentine humanist Piero Vettori, the erudite bibliophile Gian Vicenzo Pinelli, and (at the end of Mercuriale's career) Galileo.[31] His medical correspondents were also numerous. No printed collection of medical letters designated as such appeared under Mercuriale's name, but his collected "medical responses and consultations" certainly support the contention of those sixteenth-century medical authors who claimed that *consilia* were essentially indistinguishable from medical letters, since the great majority of his *consultationes* take the form of advice written to a medical colleague about the case of an individual patient (in a few instances, Mercuriale's advice is offered directly to the patient concerned). One striking example is a *consultatio* regarding a female patient, addressed in friendly terms to the Jewish physician Abraham Portaleone.[32] The first volume of Mercuriale's collected *consultationes*, also edited by Michele Colombo, appeared in 1587; subsequent editions, the last of which appeared in 1624, added three more volumes.[33]

But only a very limited idea of the extent and nature of Mercuriale's correspondence with medical colleagues north of the Alps can be gained from the advice for individual patients that fills the editions of his *Consul-*

tationes. Conceivably, the selection of material for publication in these collections involved an element of prudence—whether Mercuriale's, that of his students who prepared the successive editions, or that of his various Venetian publishers (the first volume of Mercuriale's *Consultationes* to be published was dedicated to Cardinal Enrico Gaetani, papal legate at Bologna, and carries both a papal and an imperial privilege). In the first volume of Mercuriale's *Consultationes*, 11 of 112 items are directed to northern patients or their physicians: this volume includes one of Mercuriale's letters to Crato, and one of the patients is Emperor Maximilian II.[34] Mercuriale had presumably met Crato, physician to three successive emperors from 1560 to 1581, in the latter's capacity as imperial physician when, in 1573, Mercuriale was summoned to Vienna to treat Maximilian II.[35]

Later volumes include, in volume 2, a *consultatio* sent to another imperial physician—Giacomo Scutellari, physician to Rudolf II from 1587 to 1590—with advice for treatment of the imperial chamberlain's eye affliction, and, in volume 4, advice addressed to the physician Paul Weinhart on treatment of the pain in the side and other health problems of Archduke Charles of Austria.[36] After the first volume, the proportion of northern addressees decreases: they constitute only 6 of 106 items in the second volume, and a similarly small ratio in the third. Mercuriale's collected *consultationes* thus include several items that highlight his connections with patients at the imperial court, but they reveal relatively little about his theoretical and other discussions with northern physicians. In all four volumes, patients, by and large, are not mentioned by name, unless they are persons of exceptional distinction (as in the instances just indicated). In the first three volumes, the same is true of physicians: among the few named in volume 1 are Lukas Stenglin, an anti-Paracelsian from Augsburg, and Thomas Mermann, the *medicus* of the Duke of Bavaria.[37]

By contrast, in volume 4, physician addressees, most of them relatively obscure, are given for the great majority of *consilia*. Possibly this part of the collection, in which the individual items are labeled *consilia* rather than *consultationes* and which appeared in print for the first time in 1604 (near the end of Mercuriale's life), may have undergone less editorial intervention than the earlier sections. In this final volume the physicians of some distinction, or occupying important positions (not the same thing), to whom *consilia* are addressed are Italian: the anatomist Girolamo Fabrizi d'Acquapendente, the pioneer of forensic medicine Giovanni Battista Codronchi, and the papal physician Giacomo Bonaventura.[38] To Codronchi, the author of a treatise endorsing the reality of witchcraft, Mercuriale wrote an assurance that there was no physiological or anatomical basis for the superstition that the cartilage of the xiphoid process at the base of the sternum was the seat of the soul and could be displaced by demonic magic.[39]

Mercuriale's letter to Bonaventura, the papal physician, is not a *consilium* but a defense of the style of his *consultationes/consilia*—and perhaps, implicitly, of the extent of his northern connections (even in the very restricted form in which they appear in his published *Consultationes*). In this letter Mercuriale again called attention to his summons from Maximilian II, noted the dissemination "in almost all regions of Europe" of the innumerable *consilia* he had been called upon (or, as he said, "compelled") to write in almost fifty years of medical teaching and practice, and defended their brevity against critics who called for a wider range of citations and for more theoretical argument. *Consilia*, Mercuriale held, should be succinct and to the point, and he cited Celsus, "who wrote that the sick are cured not by words or eloquence but by remedies," in support of concise medical advice. Mercuriale defended the absence of extended theoretical arguments in his own *consilia* as a deliberate choice, remarking that "anyone who read my various commentaries and who heard me publicly teaching in various Italian universities easily knew that I could discourse without difficulty about diseases, about the nature of things, and other similar matters," an observation amply borne out by the content of his commentaries.[40]

Mercuriale also exchanged other medical letters, different in character from the contents of his collected *Consultationes*, with physicians outside Italy. These correspondents included—as well as Crato and Monau—Theodore Zwinger at Basel, the Calvinist physician, polymath, and encyclopedic source of information on history, travel, geography, and antiquities. Unlike Mercuriale's *Consultationes*, which were frequently reprinted by Venetian presses, this correspondence was not published in Italy. Mercuriale's letters to Zwinger remain in manuscript, and his correspondence with

Crato, Monau, and others was published only in Scholz's collections. For Zwinger, collecting a vast range of correspondence—the library of Basel University has well over two thousand letters addressed to him—was doubtless one more way of assembling encyclopedic information. For Mercuriale, an exchange of letters with learned physicians in the north that carefully avoided any hints of religious or philosophical controversy enabled him to stay in touch with intellectual life outside Italy, notwithstanding the intensifying religious censorship of the last forty years of the sixteenth century.[41]

Mercuriale corresponded with Zwinger for at least fifteen years. More than ninety of his letters to Zwinger survive in Basel, dated from 1573 to 1588; the correspondence ceased only with Zwinger's death. For the most part these letters are short and personal; many of them concern practical arrangements or requests for information that reflect Mercuriale's attempts to advance his intellectual interests and agenda. As an avid book collector, he constantly inquired after and tried to obtain copies of scholarly books published in northern Europe.[42] His successful effort to arrange publication of Basel editions of some of his own medical works, as well as of philosophical and scientific writings of other Italians, relied on Zwinger's close connection with Pietro Perna, the exiled Italian printer in Basel. The last work that Zwinger received from Mercuriale was the recently published first volume of the latter's collected *Consultationes*; Mercuriale's accompanying letter described the book as a gift for Zwinger and did not explicitly propose a Basel edition. Nevertheless, this work, too, appeared from the press of Conrad Waldkirch, Perna's son-in-law and successor, in 1588.[43] Mercuriale's warm reception of Zwinger's recommendations of students who came from Basel to Padua ensured that some of those students, in turn, were the means of maintaining further contacts with Basel. And through Zwinger, Mercuriale sought information about other northern learned physicians and their projects. When he heard from Zwinger that Erastus was about to return to Basel from Heidelberg, Mercuriale replied: "because I greatly delight in his learning and writings I cannot but rejoice greatly in his coming to you, from which place [Basel] it will be easier to write and receive letters."[44]

Different in character both from his collected *Consultationes* and from the letters to Zwinger are the expositions of and arguments about issues in medicine and natural philosophy that fill the samples of Mercuriale's cor-

One of Girolamo Mercuriale's (1530–1606) many letters to Theodor Zwinger, April 17, 1575, Basel University Library, Frey-Gryn Manuscript II.4.173r, courtesy of the Basel University Library. In this letter, as in a number of others, Mercuriale arranges for the publication of one his own works in Basel, while simultaneously asking Zwinger to send him books published in the north. Here, Mercuriale informs Zwinger that he relies on him to supervise the publication of his, Mercuriale's, *Lectiones variae* by the Basel printer (and Italian religious exile) Pietro Perna. Mercuriale adds that he no longer needs a copy of Conrad Gesner's *Bibliotheca universalis*, as he had already obtained one from another source; instead he would like to have Lilio Gregorio Giraldi's (1479–1552) dialogues on Greek and Latin poets (Basel, 1545).

respondence with Crato, Monau, and others in Scholz's collections. Mercuriale was already engaged in medical correspondence outside Italy within two years of his appointment to the Padua professorship, even before the beginning of his correspondence with Zwinger. The earliest of Mercuriale's letters published by Scholz, written in the first part of 1571, is a reply to a letter from Wenceslaus Raphanus, a physician from Breslau and an associate of the celebrated Hungarian humanist, former bishop, and probable Socinian Andreas Dudith.[45] Raphanus sought Mercuriale's opinion on three topics: the nature of a current epidemic, the effect of dysentery on the liver, and the validity of love philtres. Mercuriale's response apparently satisfied Raphanus only with respect to the epidemic and dysentery. He remained in strong disagreement with Mercuriale's opinion that love philtres "do not exist and should not be considered by prudent medical practitioners" and that the best love philtre was similarity of manners, education, and age, combined with honorable behavior. Consequently, in a second letter Mercuriale sharply reproved Raphanus for his continued faith in love philtres, equated "pharmaca" with poisons, referred darkly to the probable (though not proven) reality of witchcraft, and was frankly incredulous about an anecdote with which Raphanus had tried to support his position.[46] Raphanus in turn passed Mercuriale's letter on philtres on to Dudith, who still remembered it, and sought additional information on the subject, more than ten years later.[47]

Letters published in Scholz's collections include samples of Mercuriale's correspondence with several physicians holding appointments at the Habsburg imperial court at Vienna and, later, at Prague. One of them was the botanist Rembert Dodoens, who was physician to the Emperor Maximilian II for a few years in the 1570s. Mercuriale debated the causes of kidney stones with Dodoens, countering Dodoens's theory that they resulted from consuming sugar by presenting examples of peoples (Neapolitans, Sicilians, and Turks) who ate a lot of sugar without suffering this affliction.[48] The few letters exchanged between Mercuriale and Crato were also entirely medical in content. Their subject was—as it had been for Erastus and the Paduans—the nature and transmission of disease. When the two physicians corresponded in 1580, they did so in the context of a current epidemic of "catarrh and fever" that Mercuriale described as "severe but not fatal," which was spreading "almost throughout the whole of Europe" and was rapidly communicable from one person to another.[49] The particular point at issue was whether contagious diseases necessarily involved some kind of putrefaction, as Crato maintained and Mercuriale denied. The discussion of the nature and transmission of disease in these letters certainly substantiates Vivian Nutton's point that Fracastoro's ideas about contagion, far from being ignored by his contemporaries, were in one form or another widely discussed but readily absorbed into the existing matrix of theories about the causes of disease.[50]

In the following years, from 1581 to 1584, Mercuriale continued an exchange of letters with Peter Monau, who, on Crato's recommendation, had briefly become the latter's successor as one of the physicians of Emperor Rudolf II.[51] Monau seems to have started the correspondence by writing to Mercuriale with further arguments in support of Crato's view of putrefaction and contagion. But he also used the same letter to shift the focus of their exchanges away from medicine and into the realm of natural history, specifically mineralogy. In the sixteenth-century context, there was nothing unusual in the combination of medicine with the study of other *naturalia*, as the careers of many individuals more celebrated than Monau— among them Georg Agricola and Conrad Gesner—show. Yet Monau raised a question that was especially problematic for physicians, because it potentially entailed a radical critique of the Galenic version of matter theory. His inquiry concerned the properties of lead: since, according to Galen, the elements air and water (with their qualities of frigidity and humidity) predominated in the temperament of lead, how was it that lead was so heavy, compact, and dense? Why was not earth, in which supposedly the heaviest element predominated, heavier than lead? Would not any presence of air in lead require a space, as in sponges or pumice? And did not the same objections regarding weight apply in relation to other minerals: for example, to the traditional ascription of an airy and watery temperament to mercury?[52]

In reply, Mercuriale brushed aside Galen as simply wrong on this point and urged Monau to follow "Aristotle, sense, and reason" in recognizing that the elementary mixture in lead and other metals consisted of earth and water, not air and water. Mercuriale also poured scorn on an unnamed author who believed that the air in lead made it increase or decrease in weight or size at different times and who offered as evidence the assertion that this occurred in leaden chains securing statues. This was a hit at Pietro Andrea

Mattioli, who had included both the claim and the "evidence" about chains on statues in his famous commentary on Dioscorides.[53] Mercuriale, dotting the i's, added that apparent changes in leaden roof tiles were the result of weathering, not of changing amounts of elementary air in the metal.

Monau waited almost two years to write back, but the response he finally sent defended Galen with vigor. After expressing his own preference for the opinions of Galen, Avicenna, Albertus Magnus, and unnamed "recentiores" over those of Aristotle, he asserted that the views of all of them, including *chymici* who assigned the principle of metals to sulfur and mercury (presumably Paracelsian physicians or other alchemists), could be reconciled, just as the views of philosophers and physicians harmonized regarding the four elements and the humors in the composition of the human body. Monau tried to add further experiential evidence for the variability of lead in size and weight by claiming that merchants left their supplies of lead outside in the rain in daylight, in order to raise the air and water content and thereby secure a higher price. And starting from a conventional Aristotelian account of the formation of metals underground, he moved on to invoke traditional "experiences" that in his view demonstrated natural augmentation of nonliving matter and thus supported his conviction that given the right conditions, "some extrinsic augmentation" of lead could take place. The experiences in question were the revival of (apparently dead) perennial plants after winter, the supposed resuscitation of dead insects and small animals by heat, and the alleged continued growth of hair and nails on corpses.[54] Moreover, Monau's response to Mercuriale's exhortation to pay attention to "Aristotle, sense, and reason" was to point out supposed inconsistencies in Aristotle's account of the process of melting and to raise yet another question about the properties of lead: why, Monau wondered, did not the great leaden cauldrons in which salt was rendered in German mines (for example, in Luneburg, Saxony's famous salt-mining center) melt from the heat of the fire beneath them?[55]

The discussion, or rather disputation, between Monau and Mercuriale about metals rumbled on through several more letters. By the end of 1583 Mercuriale was excusing himself for not responding more quickly to further arguments about lead on the grounds of his many professional obligations—"what with visiting the sick, giving public lectures in the university, and my own studies, I scarcely have time to cut my nails."[56] Monau did not give up on the subject of the leaden cauldrons used in processing salt, suggesting that perhaps the properties of the salt water itself prevented the containers from melting.

But Monau also turned back to medical topics. He evidently made an effort to obtain and read Mercuriale's publications as soon as possible, and promptly thereafter sent the author critiques of and questions about them. Thus Monau's letter to Mercuriale of January 1584 contains comments on two of Mercuriale's short treatises on specific medical topics. The first was his work on diseases of children, published the previous year; Monau called into question its recommendations regarding purgatives for infants and informed Mercuriale that a parasitic infection not mentioned in the book was prevalent among children in Germany. The second was Mercuriale's recently published treatise on poisons, in response to which Monau raised both philosophical and medical objections to Mercuriale's fundamental definition of poison.[57] But Monau also circulated information about Mercuriale's works among German colleagues and sought information about his future publications. In a letter of February 1584, he reported to Mercuriale that he had received two copies of the Paduan's *Censura Hippocratis*—first published in 1583, this was one of the earliest attempts to evaluate the authenticity of the works of the Hippocratic corpus—and had soon passed them on to other learned men in Germany. In the same letter he inquired eagerly about Mercuriale's projected edition of Hippocrates, expressing the hope that it would include the Greek text as well as a Latin version. But Monau also added some more criticisms of the work on poisons and attempted to initiate a new thread of discussion about the function of the spleen.[58]

In some respects these last two letters of Monau's, with their avidity for new books and requests for information about scholarly projects, recall Mercuriale's much longer, and unpublished, series of letters to Zwinger. For Monau, too, letters sent abroad may have served as a way of keeping in touch with intellectual life elsewhere in Europe. But despite many expressions of courtesy on both sides, the correspondence between Mercuriale and Monau always centered on disputation. Authors, alas, do not always welcome prompt criticisms of their work from younger colleagues, and Mercuriale eventually had had enough. His last published response to Monau was an exceedingly sharp rebuttal of all of Monau's critiques of his books as well of Monau's natural philosophical ideas about lead and physiological theories about the spleen.[59]

Across the Religious Divide: The Galenist Response to Paracelsianism

With the exception of Mermann at the ducal court of Bavaria, whose rulers were enthusiastic proponents of the Catholic cause, and the possible exception of Dodoens, the well-known physicians in the German lands and central Europe with whom the professors from Padua corresponded in the 1570s and '80s were, to a greater or lesser extent, committed to—with varying degrees of openness—one version or another of reformed religious beliefs. Erastus was active both as a theologian and as a professor of medicine in reformed Heidelberg. Crato, who had studied with Luther and was close to Melanchthon, seems to have sought eirenic religious compromise at the court of Emperor Maximilian II; Monau, who completed his education at Basel, was probably a crypto-Calvinist. Doubtless medical training and interests, and in several cases earlier personal contacts or common acquaintances at Padua or elsewhere, provided sufficient common ground to make correspondence possible. But something else may also have appealed to the professors from Padua: the stance of Erastus and Crato as anti-Paracelsians.[60]

At a time when Paracelsus's ideas were just beginning to be disseminated throughout Europe in Latin translations of his works, Erastus and Crato emerged as strong defenders of what they perceived to be true Galenist orthodoxy in medicine. Furthermore, not only Catholic religious authorities but also some men of learning who openly adhered or quietly inclined to reformed religion (Erastus and Crato among them) perceived Paracelsus as unorthodox in religion as well as in medicine.[61] By 1573 Mercuriale was keenly interested in Erastus's anti-Paracelsian writings, as some of his letters to Zwinger make evident. It is unclear to what extent Mercuriale was ever fully aware of Zwinger's own complex attitude toward Paracelsus, which, as Carlos Gilly has shown, moved in the early 1570s from rejection to an evaluation that combined criticism with an appreciation of Hippocratic and empirical elements in Paracelsian medicine. Mercuriale himself owned several works of Paracelsus (some but not all in editions printed by Perna) as well as those of Erastus, but there seems no doubt that his own position lay within the Galenic mainstream. In 1574, in the first of his letters to Capodivacca, Erastus himself specifically called the attention of the Paduans to his anti-Paracelsian writings.[62]

In Italy, where the importation of some Paracelsian books was prohibited, the penetration of Paracelsian books, ideas, and remedies may have been somewhat slower than in other parts of Europe, but it was certainly beginning by the 1570s.[63] In this context, one remaining letter to Crato from an Italian physician, published by Scholz, calls for attention. In March 1585, Girolamo Donzellini wrote from Venice to assure Crato that he shared the latter's hostile opinion of Paracelsians, and to inform him that he was sending him a copy of a pseudonymous treatise that he, Donzellini, had written against a work in Italian titled *The Scourge of Rational* [i.e., Galenist] *Medicine* by a "chymical empiric." In the same letter Donzellini insisted that he did not think such remedies were safe (but went on to give a couple of recipes for them). As for books by Paracelsus, "At one time I had all the books of Paracelsus by me, but since nothing useful came from reading them, I exchanged them for others, nor did I want to have in my library such books which taught nothing."[64] The treatise against Galenic medicine Donzellini referred to was the work of his onetime friend, the Italian Paracelsian Tommaso Zefiriele Bovio. But in all probability it was Donzellini—whose contacts during some years spent in Basel and Germany included Pietro Perna, the Basel publisher of many Paracelsian works—who introduced Bovio to the works of Paracelsus in the first place, and who may have played a major part in introducing Paracelsian (and other forbidden) books into Italy.[65] Donzellini's letter to Crato was evidently an attempt to extricate himself from a net that by 1585 was finally closing. He had already endured several previous investigations by the Inquisition, recanted what he claimed were only briefly held unorthodox views, and undergone a period of imprisonment. At last, in 1587, as is well known, he was executed as a relapsed heretic by being drowned in Venice's lagoon.[66]

Donzellini's fate was not due solely to his flirtation with Paracelsian medicine. As Richard Palmer has shown, he had a long history of dangerous contacts with northern supporters of religious reform, injudiciously frank letters about his religious beliefs, importation and ownership of banned books, previous accusations, and abjurations; and he seems to have been exceptionally reckless.[67] Moreover, the 1580s saw the most intense phase of a period of religious repression in Venice.[68] As a writer of medical letters, then, Donzellini stands in sharp contrast to the handful of academic physicians who managed to correspond across the religious and political boundaries of sixteenth-century Europe without harm to position or reputation. The letters on strictly medical and natural philosophical topics that Mercuriale and other professors at Padua exchanged with colleagues in the north who were like-minded orthodox Galenists (and hence strongly anti-Paracelsian), especially if the latter were also associated with the imperial court, did not necessarily entail the same risks. Yet it should be noted again that Mercuriale's letters to Zwinger remain unpublished; moreover, after the publication of Massa's collection in Venice at midcentury, discursive letters between Paduan professors and northerners (as distinct from the brief items of advice for a handful of distinguished patients contained in Mercuriale's *Consultationes*) were printed only in collections published in the German lands. In the second half of the sixteenth century, the medical republic of letters was surely in some sense a reality, but not one free of the constraints of the age. In contrast to the more far-flung communications discussed here, the following chapters turn to examples of medical letter collections by authors whose correspondence networks did not, for the most part, extend across both northern and southern Europe or involve the crossing of cultural or religious boundaries.

EFFIGIES
CL. V. D. IOANNIS
LANGII LEMBERGII
Medici, Archiatri Palatini
Electoralis.

Archiatrum facit Heidelberga: Sophum facit ante
Lipsia: sed Medicum Felsina docta bonum.
N. R.

Johann Lange (1485–1565), portrait frontispiece from his *Epistolarum medicinalium volumen tripartitum, denuo recognitum et dimidia sui parte auctum* (Hanau: typis Wechelianis, apud Claudium Marnium & haeredes Ioannis Aubrii, 1605), courtesy of the New York Academy of Medicine Library. The portrait was first published in Lange's *Secunda epistolarum medicinalium miscellanea, rara et varia eruditione referta, non medicinae modo, sed cunctis naturalis historiae studiosis plurimum profutura* (Basel: [ex officina Nicolai Brylingeri, expensis Ioannis Oporini], 1560), where it is accompanied by a caption stating that it represents Lange in 1558.

CHAPTER TWO

The Court Physician Johann Lange and His *Epistolae Medicinales*

WRITING IN 1589, Nicholas Reusner situated the published *Episto-lae medicinales* of Johann Lange, longtime court physician at the court of the Elector Palatine in Heidelberg, as follows:

> Among his works is this miscellany of *Epistolae medicinales*, rich in varied and rare erudition as also in authoritative explanation of all kinds of very worthy things . . . Indeed I am not ignorant that a number of people have composed outstanding works in this genre by writing and publishing similar collections of *epistolae medicinales*. Especially deserving of commendation are Giovanni Manardo, Luigi Mondello, Giovanni Battista Teodosi, Pietro Andrea Mattioli, Nicolò Massa, Vettor Trincavella, and if there are others besides those named I do not have their names. But these are all Italians; unless I am mistaken, Lange is the only German.[1]

Reusner was Lange's distant relative; his words occur in the preface he wrote for a posthumous edition of Lange's *Epistolae* prepared by yet another relative, Georg Wirth, for a time a court physician to Charles V and Philip II and Lange's confidant and heir. Lange's German world was, as Reusner's preface makes clear, shaped in important ways by his membership in a family network of physicians and jurists. After arguing, with numerous historical examples, ranging from antiquity to the recent past, that "really there can be a perpetual possession of that art [of medicine] in outstanding families, as was recorded of the Asclepiads," and that "[t]his [family tradition] was also accustomed to be noted in other arts and disciplines," Reusner went on to specify the names and relationships of distinguished physicians and jurists in the Wirth-Reusner-Lange family network: an earlier Georg Wirth, physician to Louis II, king of Hungary and Bohemia; his brother Peter Wirth, theologian and dean of the faculty of arts at the University of Leipzig; Michael Wirth, a jurist who was dean and rector of the University of Leipzig and later chancellor of Saxony; Bartholomaeus Reusner and Hieronymus Reusner, physicians; and Christoph Reusner, jurist.[2] No doubt Reusner's insistence on Lange's German identity reflects familial pride and perhaps also a reaction against dismissive treatment of northern Europe culture by some Renaissance Italians. Nonetheless, he was essentially correct both in associating Lange's work with that of the early and mid-sixteenth-century Italian pioneers of the genre of published collections of *epistolae medicinales* and in emphasizing the significance of Lange's different cultural and regional context.

The contents of Lange's "medical letters" (as will become apparent, not all letters, and not all strictly medical) range across an array of topics and formats wide enough for Reusner to claim that they were useful for the Republic of Letters in general and for all students of literature, as well as for medicine and natural philosophy. But collectively they illustrate four major elements, frequently interrelated, of Lange's formation and career: (1) a professional identity partially shaped by medical humanism, (2) the multifarious duties of his long service as court physician to successive rulers of the Rhineland Palatinate, (3) the intellectual interests of a local and regional circle of men of learning (physicians and others), and (4) contemporary and local social and religious controversy. At different times historians have noted all four of these aspects of Lange's life and work, though not always viewing them with equal attention and in relation to each other.[3] The objective of this chapter, after a brief biographical summary, is to reexamine some of the letters in the light of these four themes and, especially, their interaction.[4]

Lange (1485–1565), who was born in Silesia, followed other members of his family to the University of Leipzig, where he matriculated in 1508, received his master's degree in 1514, and taught arts for several years. At Leipzig, where he was among the first group of humanists to teach at that university, he lectured on Cicero, Pliny, and the pseudo-Aristotelian *De mundo*.[5] According to his own account, Lange also lectured on pseudo-

Proclus, *De sphaera*, and taught cosmography at Leipzig. The long discourse on the voyages of exploration in the New World and the East in one
of his letters, which also includes travel advice for a journey to Calicut,
although apparently dating from the 1540s, perhaps echoes some of his
cosmographical teaching from the Leipzig years.[6] Lange also delivered the
concluding oration at the famous Leipzig disputation in 1519 between the
Catholic theologian Johann Eck and the reformers Luther and Carlstadt.[7]
Thereafter, like many other northerners, Lange went to Italy for medical
training and, after studies at Bologna, graduated with his degree in medicine from Pisa in 1522 or 1523.[8] It is clear that he used his time in Italy to
engage fully with current intellectual interests and, wherever possible, to
seek out significant figures involved in them. Lange's professor of medicine
at Bologna was the obscure Ludovicus de Leonibus, but a trip to Ferrara
enabled him to meet the exemplar of medical humanists, Niccolò Leoniceno. Moreover, Lange also attended lectures by Pomponazzi, whom he
later described as "my teacher of philosophy at Bologna" (*meus Bononiae in
philosophia praeceptor*).[9] He improved his knowledge of Greek language
and literature by study with Pietro Ipsilla of Egina, the professor of Greek
at the University of Bologna, who had also been the Greek teacher of Pope
Leo X (Giovanni de' Medici).[10] Summer vacation provided Lange with opportunity for a visit to the University of Padua and to Venice, where he
explored Venetian fish cookery. Yet another trip in student days took him
to visit Gianfrancesco Pico.[11]

Shortly after his return to Germany, Lange became court physician to
the ruler of the Rhineland Palatinate, a position that he held, under successive rulers, until his death—a span of more than forty years. He spent
most of the rest of his life in Heidelberg, although he also traveled quite
widely in the early years of his court service. He left a diary of a journey in
1526 in which he accompanied the future Elector Palatine Friedrich II
through the German lands, France, and the Iberian Peninsula to Granada
and back; Lange was also part of Friedrich's entourage on two military
expeditions against the Turks. In the early 1540s, Lange seems to have returned to Italy for a visit, during which he claimed to have discussed travel
to and imports from the New World and the East with physicians, travelers, merchants, and naturalists in Venice.[12]

Lange published his principal medical works—a dialogue titled *Medicum de republica symposium* and two *miscellaneae* of *epistolae medicinales*—late in life. Both the dialogue and the first of the two collections of *epistolae medicinales* appeared in 1554, when he was already sixty-nine years old. Lange's first set of *epistolae medicinales*, containing eighty-three items, was published by Oporinus in Basel and evidently rapidly attracted the attention of other learned editors and readers. Its contents included a group of eleven of Lange's letters on surgical topics that must have favorably impressed Conrad Gesner, for he included them in his collection of works on surgery published in Zurich the following year. In 1556, the entire contents of Lange's first set of *epistolae medicinales* were reprinted in the anthology of medical letter collections by different authors titled *Epistolae medicinales diversorum authorum*, published in Lyon. It seems likely that Reusner had this volume in mind when he associated Lange with Italian writers of medical letters, since the anthology's other authors—Manardo, Mondella, Massa, and Teodosi—were all Italian and correspond, with two exceptions, to Reusner's own list of such authors. In 1560 Lange followed up his first collection of *epistolae medicinales* with a second, also published in Basel, containing sixty-one items. After Lange's death members of his extended family undertook the task of editing his work for further publication in new and enlarged versions. Two posthumous editions of the *epistolae medicinales* in three books appeared in Frankfurt, in 1589 and 1605. In these, the two collections published in Lange's lifetime were redesignated books 1 and 2, respectively; book 3 consisted of a reissue of the dialogue *Medicum de republica symposium* and five previously unpublished items. In addition, at various later dates in the sixteenth and early seventeenth century, a few individual letters from Lange's collections were reprinted with works of other authors.[13]

Medical Humanism and Professional Identity

In his first collection of letters, Lange informed the reader that when he left teaching arts to study medicine, "lest I should seem to have abandoned source springs for rivulets [*fontes per rivulos*] I pored over works of divine Hippocrates and his faithful interpreter Galen day and night."[14] Undoubtedly he was a medical humanist in the sense of one who sought to turn to original Greek medical writings and avoid medieval translations, summaries, and interpretations. His letters are filled with citations of classical medical authorities and exhortations to follow the example of the ancients by observing symptoms closely and reasoning about causes, an endeavor impossible, in his view, without natural philosophy.[15] A long discussion of the causes of pain resolves into a demonstration that medieval Arab authors and Latin scholastics—Avicenna, Averroës, and Pietro d'Abano—all failed to understand Galen's explanation correctly.[16] But Lange's break with the standard medical authorities of earlier generations was far from complete. Like many of his contemporaries, he cited Avicenna frequently and often with respect, and, occasionally, Latin scholastics; more significantly, he explicitly advocated retaining a place for the "medicine of the Arabs" in a reformed university medical curriculum.[17] Lange's humanism also presented other distinctive features besides deep commitment to Greek medical authorities. One, several times noted by historians, is the aggressive use he made of claims for the superiority of humanist medical learning to denounce not only specific diagnostic and therapeutic methods that he regarded as without rational basis but also whole categories of practitioners outside the circle of university-trained physicians. Of course, such denunciations were by no means peculiar to Lange, but his attacks stand out for their ferocity, repetitiveness, and focus. Among medical procedures, Lange especially derided uroscopy; and he poured scorn on military surgeons who treated a fever that caused darkening of the tongue by applying red cloth to that part of the body.[18]

Lange shared the hostility of Manardo, the author of one of the earliest and most influential published collections of *epistolae medicinales*, toward the use of divinatory astrology in medicine and severely condemned the technical ignorance and medical harmfulness of contemporary astrological practitioners—in Lange's words, "a crowd of dealers in trashy finery" (*nugigerulum astrologorum vulgus*). He attributed some of their errors to reliance on a methodology derived from the Arabs. The most extended discussion of astrology in his letters praises the usefulness for physicians of knowledge of star risings and settings, climate, and seasons, citing Hippocrates's *Airs Waters Places*, but condemns reliance on planetary houses and aspects in selecting days either for medical treatment or for activities and decisions of daily life. In the very next letter he followed up with a case history—the sad and untimely death of Peter Wirth after a phlebotomy recommended by a *medicus* was delayed for three days for astrological reasons, on the advice of an apostate monk. Denunciations of divinatory astrology also occur in other letters by Lange.[19]

His opinion of alchemy seems to have been not much different from his opinion of astrology, though his writings on the subject were somewhat more ambiguous—possibly on account of the enthusiasm of Friedrich II's nephew and successor, the Elector Ottheinrich, for alchemical practice. Lange devoted one lengthy letter to a minihistory of the discipline, accompanied by a critical evaluation of various alchemical procedures, showing some knowledge of their vocabulary and techniques. According to Lange, alchemy, like astrology, was an art with very insecure foundations on which nothing lasting could be built, and popular interest in alchemy was based on greed. Yet he also attributed its origins to secret wisdom that Moses took from Egypt and acknowledged that the techniques of the chymists had produced many medicaments much used in the cure of diseases. He ended the letter with a warning against the disappointments liable to await those who attempted alchemical practice.[20] Lange concluded another letter, addressed to a relative of whom he was fond, with a strong warning against putting any trust in "those wandering medical practitioners, who promise undiminished youth . . . and immortality from their potable gold and elixir."[21]

But Lange's was a world of medical pluralism, even if not, in the strict sense, a "medical marketplace," with a rich vernacular medical culture (in which he, too, participated).[22] In the mid-sixteenth-century Palatinate, in all likelihood informally trained practitioners of various types greatly outnumbered graduate physicians, let alone medical humanists trained in the Italian schools. Lange's repeated denunciations of the deficiencies of medical and surgical practitioners other than graduate physicians clearly had a social as well as an intellectual, and probably also an economic, basis. But it is equally clear that in some instances religious issues shaped his views—a topic dealt with in more detail below.

Lange's humanism, like that of many other early modern learned physicians, was by no means narrowly medical, as his pattern of citation, his use of literary forms, and the topics he addressed all attest. Thus, although Galen and Hippocrates are undoubtedly the authors most frequently mentioned in his *Epistolae medicinales*, citations of Plato are also especially numerous; in addition, Lange referred to Greek historians and literary

authors, among them Herodotus, Diodorus Siculus, Athenaeus, and Aeschylus. Moreover, it seems likely that he paid a good deal of attention to
the literary presentation of his work. Presumably the contents of at least his
first *miscellanea* had been assembled over a period of years, but in both his
collections, unlike some other published volumes of *epistolae medicinales*,
the individual letters are undated.[23] Instead, many of Lange's letters are
gathered into loose subject groups. This arrangement is by no means system-
atic or complete, but, while not intended as a comprehensive list of all such
clusters, for example, in the 1605 version of the *Epistolae*, book 1 (= *Miscellanea* 1554), letters 3–10 concern the ignorance and errors of surgeons; 16–25
discuss specific disease conditions and their treatment; 27–31 are on drinking;
33–36 and 38, on magic and astrology; 45–47, on urinary problems; 50–52, on
ancient baths and gymnasia; 55–62, on food; and 63–76, on plants, drugs,
and poisons. In book 2 (= *Miscellanea* 1560), letters 5–12 are on sexual problems; letters 18–22, on medicaments. Hence, although both *Miscellaneae*
contain letters reporting episodes from early in Lange's career, there is seldom any way of knowing either when such items were originally composed
or how much reworking preceded publication. Some revision—or even
literary fiction—is suggested by the use of pseudonyms for many of the
addressees in the first collection.

Moreover, despite the title *Epistolae medicinales*, a number of items are
only nominally letters: after the briefest of epistolary salutations, they contain a dialogue. Whatever their basis in remembered conversations, these
are literary compositions. Whether Lange should be accounted a Platonist
in philosophy is unclear; but his interest in the Platonic dialogues, in Plutarch (another frequently cited author),[24] and in Athenaeus may surely be
connected with his liking for the dialogue form. Furthermore, a number
of Lange's dialogues are symposia in the original sense—placed in the setting, fictitious or not, of a convivial dinner party. Thus several letters report on a series of dinners during Lange's later return to Venice in which
he, his cousin Georg Wirth, and Venetian physicians purportedly discussed
with Geraldus, a young graduate in medicine recently returned from a
voyage to the East, and a group of *medici herbarii* just back from Egypt
such subjects as "the exotic medicaments of the Arabs" and what Eastern
plants and fruits were known to ancient authors, to the accompaniment of
jests with the maidservant and comments on the menu.[25]

In subject matter the letters (with or without embedded dialogue) range over not only medical problems and questions but also a variety of contemporary intellectual issues, among them antiquarianism, magic and demonology, the wonders of ancient Egypt, and natural history. Lange's views on magic and demonology and on ancient Egypt were largely shaped by his religious views. Letters on Roman baths, on the meaning of terms in the Hippocratic corpus for various forms of exercise, on the origins of alchemy, and, in a lighter vein, on the varieties of material from which the ancients made drinking cups give an idea of the scope of his antiquarian interests.[26] In these epistles, with their freight of learned citation, Lange's antiquarianism looks almost entirely bookish and focused solely on Greek and Roman antiquity. But he had some personal acquaintance with physical remains, both of classical antiquity and of medieval northern Europe: his travels included seeing the ruins of Roman baths in Granada, and he participated with interested skepticism in uncovering the supposed tomb of Roland.[27]

In another context, Lange gave antiquarianism contemporary relevance in his comparison of Ottheinrich, Friedrich II's successor and the chief collector of the famous Palatinate library, with founders of libraries and famous book collectors from Ptolemy Philadelphus to Giovanni Pico della Mirandola. Lange embedded the comparison in an overview of library history that was much shorter and considerably less detailed than the account that Justus Lipsius would publish some sixty years later, but that nonetheless assembled references to libraries from more than fifteen ancient authors. Ottheinrich was a benefactor to whom Lange had every reason to be grateful: on his accession as Elector Palatine in 1556, he had relieved Lange of many duties, appointed him a counselor, and turned to him for advice on university reform. Yet for all its courtly flourish, Lange's compliment to Ottheinrich—an extraordinarily lavish patron of arts and sciences of many kinds as well as an avid, and sometimes ruthless, book collector—was perhaps not unjustified.[28]

Interest in natural history, seen primarily from the standpoint of *materia medica*, was more central to Lange's medical learning. Like many other contemporary physicians, he attributed great significance to the correct identification of medicinal plants named by ancient authors, which he saw as an essential tool for avoiding substitutions made by unscrupulous or unwitting pharmacists. Although he lived in a period of important direct botanical investigation and knew the work of some of its practitioners, most of his own treatment of the subject was semantic and philological; he devoted much effort to disentangling plant names in Galen, Theophrastus, Plutarch, Homer, and Aetius and suggesting correspondences among them.[29] Nevertheless, on at least one occasion he embarked on a detailed description, suggestive of personal knowledge, of a plant growing in the forests and meadows of the Palatinate:

> In March in the Hercynian forest, near Kreuznach and Ingelheim, towns of the Palatinate beyond the Rhine, between Bingen and Mainz, in wild places, there grows a plant nine inches high, usually having three leafy stalks, with leaves [like those] of artemisia; in the head of each leaf are single flowers . . . like camomile flowers, which after they have fallen are succeeded by a pod filled with black seeds like those of common white dittany. The rather numerous roots are black, slender, and twisting and come from a bulb at the base of the stalk. In taste and purgative power they are in all respects equal to the roots of black *veratrum* [false hellebore]; our wandering root cutters, and those of Italy too, sell them to our pharmacists and those of the Venetians as roots of black *veratrum*; nor, as I have seen, do erudite Italian physicians use them any differently. Our veterinarians use them to purge horses . . . I think this [plant] is the lesser hellebore of our Dioscorides and Theophrastus . . . and that its roots are substitutes for the black *veratrum* of the ancients.[30]

The Tasks of the Court Physician

As a growing body of scholarship has shown, distinctive responsibilities, opportunities, and, sometimes, dangers characterized the practice of medicine at Renaissance and early modern European princely and noble courts.[31] Because the outcome of medical treatment of a ruler or powerful noble and his family could have implications for a lineage and, in some instances, for a larger political world, court service presented hazards as well as opportunities. Yet court physicians could become trusted advisers who not only provided medical advice and treatment but also served in nonmedical capacities and, in so doing, acquired considerable influence.[32] Lange was only one of many court physicians whose services for a ruler encompassed both personal medical attendance and much else besides.

In writing about medicine, Lange's orientation was almost entirely toward practice—that is, disease and treatment. Apart from two letters on, respectively, different types of heat and the reconciliation of supposedly Aristotelian and Platonic definitions of *virtus informativa*, he showed almost no interest in physiological theory or in the intersection of medicine and philosophy.[33] Most of his letters on strictly medical topics are, in reality, general essays; among their subjects are diseases of the head, epidemic diseases, the correct treatment of fever, the proper care of wounds (including gunshot wounds), urinary conditions, and diseases of women. By contrast, relatively few letters give information about individual patients and cases. Lange's *Epistolae medicinales*, unlike those of some of his contemporaries, contain few disguised (or undisguised) medical *consilia* for particular individuals.[34] A somewhat larger number of items mention or comment on specific cases, whether those of his own patients or those of colleagues, but most of the descriptions are too brief to be termed case histories. Lange's medical practice and contacts at Heidelberg evidently extended beyond the court: he treated "many" during an outbreak of plague; he was called to pronounce on the obstetrical problems of the wife of the prefect of an obscure village; and he gave advice about exterminating insect pests in the Palatinate that were destroying pine trees used for timber and fuel (after a flurry of learned citations, Lange's advice boiled down to a recommendation to fumigate).[35]

Nevertheless, the prevailing impression is of a practice centered on a relatively small court circle and on contacts made through that circle. Several letters refer to Lange's treatment of various illnesses of Friedrich II and Ottheinrich; the few additional named patients included Arnoldus, "[my] friend, and counselor of our prince" (*amicus, et principis nostri a consiliis*); Theodorulus, the small son of "Patrocles [a pseudonym, like the names of many addressees in the first *Miscellanea*], [of] true glory and outstanding distinction among my patrons" (*patronorum meorum vera gloria et decus praecipuum*); the bishop and a canon of Regensberg; and Christoph Kress, orator to Charles V. Another side of court life perhaps appears in Lange's account of his successful treatment of a noble youth who, in midwinter, fell out of a window drunk and lay on the ground unconscious all night and, indeed, in the many discussions of eating and drinking scattered through the letters. One such missive, on whether wine is good for children, is addressed to Christoph Prob (the chancellor of the Palatinate) and reports a discussion at a dinner at Prob's house, when Prob's young children were present; Lange recommends that they should be allowed wine.[36] Lange's professional duties brought him into contact with those who supplied and prepared food for the court as well as with those who ate it, as on the occasions when he discussed the danger of using lead for cooking vessels and water pipes with Friedrich's cook and steward, and when he considered the safety of eating game that had been poisoned (surely rather unsportingly) by the Count Palatine's hunters.[37]

Probably the most extensive record left by Lange's actual practice is to be found not in his published letters but in the large number of medicinal recipes attributed to or amassed by him. His Latin collection of secret remedies (*De secretis remediis*), dedicated to his younger cousin and heir Georg Wirth, occupies more than half of, and is the longest item in, book 3 of the posthumous editions of the *Epistolae medicinales.* Many vernacular recipes attributed to Lange also survive in medical miscellanies among the Heidelberg Codices Palatini germanici.[38] Lange himself had tart words for compilers of large collections of medical recipes if the recipes in question—unlike his own—did not come from learned physicians:

> There are others who are called recipe collectors, who without any rational basis acquire and assemble whole volumes of recipes—enough to fill wagons—from pharmacists and chymists; these are people who, having neglected knowledge and diagnosis of diseases through the natural causes, follow only experience (*experimenta*) . . . Not the pharmacist or the chymist, but correct knowledge of the disease and its causes through the pathognomic signs indicates the correct medicaments.[39]

In the Palatinate, as elsewhere in mid-sixteenth-century Europe, confessionalism went hand in hand with the ambition of early modern territorial rulers to extend control over the institution of the university. In this context, Lange's role as personal physician to the ruler acquired a new dimension. Shortly after Ottheinrich became Elector, he undertook a major reform of the University of Heidelberg, with the goal of reshaping it into a well-regulated, and Lutheran, institution of the territorial state (Friedrich II and Ottheinrich were Lutherans; both the strong commitment of the Palatinate to Calvinism and still further extension of court influence over the university belong to the reign of Friedrich III, "the Pious" [1559–76]).[40] Among Ottheinrich's advisers on university reform was Lange, whose interest in the subject was of long standing. He had submitted a proposal for this to Friedrich II in 1545; and Lange's *Medicum de republica symposium* (published in 1554, but written early in 1547) is also, inter alia, a plea for university reform according to its author's own lights.[41] But the final version of Ottheinrich's new statutes for the medical faculty—which also incorporated contributions from Erastus, then recently appointed professor of medicine at the University of Heidelberg—followed Lange's priorities only in part. Lange's 1545 proposal had called for four chairs in medicine: one for Arabo-Latin medicine, one for surgery, and one each for lectures on Hippocrates and Galen, respectively. The statutes of 1558, going somewhat further in the direction of curricular reform on humanistic lines, provided for three professorships in what were now called therapeutics, pathology, and physiology; all three subjects were to be taught primarily from Hippocratic and Galenic texts.[42] What was adopted from Lange's proposal, almost word for word, was his draft statute both proclaiming that as the practice of medicine required knowledge of liberal arts and philosophy, practitioners must in future either be university graduates in medicine or be licensed following examination by the dean of the Heidelberg medical faculty; and banning medical practice by "Jews, parish priests, monks, also no women, or other vagrants" (*juden, pfaffen, munch, auch keinem weibsbild oder andern landfahrern*).[43]

Contacts and Correspondents

In person and as a correspondent, Lange seems to have belonged to several familiar coteries—among them a circle at the Heidelberg court, members of his extended family, and former colleagues from his time at the University of Leipzig—as well as participated in a broader regional network of correspondents. In Heidelberg itself, Lange had both patients and intellectual contacts beyond the court. One of the dialogues embedded among Lange's letters on the topic of the occult property and attractive power (*potentia attractiva*) of purgatives is prefaced by the explanation that

Emperor Antoninus Pius used to meet anyone who wanted to engage in disputation with him in evenings at the Temple of Peace, near the Roman libraries, and that this example is followed by lovers of medicine at the University of Heidelberg, who often meet for dinner at the Chilianus pharmacy, near Holy Spirit Church (where a large part of the Palatinate library was kept). In addition to Lange, one of the participants in these discussions was a medical student of whom Lange formed a high opinion, the alchemist Guillaume Rascalon, who earned a medical degree at Leipzig in 1559; another contact in the university milieu was the poet and professor of medicine Petrus Lotichius Secundus.[44] No doubt discussions of medical matters also occurred at court; the serious interest in medicine of several of the rulers (and of members of their families) whom Lange served is well documented.[45] But that some of the conversations in which Lange participated at court were less edifying than the debates at the pharmacy is perhaps suggested by his willingness to entertain bored nobles with anecdotes (including at least one very tall story) about the unnatural food cravings of pregnant women.[46]

In the years after Lange returned to Germany from Italy, his main opportunities for personal meetings with medical professionals and men of learning outside Heidelberg's court and university probably depended on the movements of Friedrich II. Thus, in addition to the earlier European travels already mentioned, Lange accompanied Friedrich to imperial Diets at Speyer in 1542 and 1544 and Augsburg in 1548. On all three occasions, he consulted with learned physicians from other parts of the Empire; at Speyer in 1542 he also had the opportunity to examine a fasting girl, whom Ferdinand of Austria had ordered brought to Speyer for investigation. From time to time a summons to attend a distinguished patient in another city allowed for interchanges with colleagues, not all of which were devoted to medical consultation: the call to Nuremberg to treat the imperial orator Christoph Kress provided the setting for a riverside discussion (real or imagined) of Greek forms of exercise and for yet another dinner party.[47]

But most of Lange's intellectual exchanges with friends, patrons, and colleagues outside Heidelberg were presumably carried on by letter. His practice of suppressing the names of many of the addressees of his letters and identifying others only by a forename or a pseudonym makes it difficult, indeed impossible, to estimate the full extent—geographical or social—of his circle of correspondents. Some of Lange's pseudonyms seem intended to imply a historical or literary reference. Thus Critobulus, the name used for the addressee of several letters, may be meant to recall either a disciple of Plato or a skillful surgeon mentioned by Pliny.[48] Yet in these missives one can catch a glimpse of three overlapping circles of friendship, patronage, and communication: family members, friends or colleagues from his years at the University of Leipzig, and a regional network of correspondents in cities of the Palatinate and Bavaria.

Most of the letters to relatives are to members of the Wirth family: Georg Wirth, imperial physician and Lange's trusted confidant and heir; Michael Wirth, jurist; and Peter Wirth, theologian and sometime dean of the Faculty of Arts of the University of Leipzig (and Lange's patron there). Lange wrote to Peter Wirth on the topic "Can the natural period of life be prolonged?" In this letter, Lange displayed some interest in, and a little knowledge of, ancient calendars and chronology, explaining away the longevity ascribed to the biblical patriarchs by an ancient device for shortening long chronologies—the assertion that the "years" in question were really months. Lange postulated a maximum natural human life span of about 100 solar years, conjoined with the standard medical teaching that the actual length of life of any individual was governed by his or her balance of innate heat and radical moisture. Medicine could help to preserve or restore that balance, so that the natural terminus might be reached. But Lange insisted that neither medicine nor any other human endeavor could prolong life beyond its natural terminus, and that the actual end of each person's existence was not fatally or irrevocably determined. A bad regimen, violence, or self-destructive behavior could cut life short, while the lives of individuals could be prolonged by divine favor. The letter ends with a warning against reliance on alchemical claims to confer "undiminished youth . . . and immortality."[49]

If "Auerbach, best of friends" (*Auerbachus amicorum optime*) is to be identified with Heinrich Stromer of Auerbach (1482–1542), physician and councilor at Leipzig who was a correspondent of Niccolò Massa, he, too, was a colleague of Lange's from university days; so was Johann Reusch, "glory of the medical faculty of the University of Leipzig" (*Lipensis medicorum academiae decus*).[50] If one leaves aside letters to family and to friends from university days at Leipzig, in the relatively few instances in which letters link the addressee with a particular place, city names—Augsburg, Nuremberg, Regensburg—suggest a regional network. Among Lange's named correspondents within this region were the physician, historian, cosmographer, and ardent Lutheran Achilles Pirmin Gasser; the physician, musician, and collector of music Georg Forster; and the short-lived and many talented Johannes Moibanus, city physician of Augsburg.[51] In contrast to various other German physicians who had completed medical studies in Italy (e.g., Peter Monau and Johann Crato), there does not seem to be any evidence in Lange's published letters to suggest that he corresponded with anyone in Italy after his return to Germany. Lange had some Swiss connections—in Basel, presumably with the printer Oporinus, who published his first collection of letters, but also with Philip Bech, physician and professor of Greek, to whom he wrote on diseases of the head; and in Zurich, with Conrad Gesner, who included some of Lange's letters in a collection of surgical treatises.[52] He also published one letter to Melanchthon, in which he urged the reformer to consider the health benefits of eating moderate amounts of cheese (Lange was trying to counter the belief that cheese was dangerously indigestible).[53]

Religion and Magic

At some point during his career Lange converted to Lutheranism, although it is not clear when. His oration concluding the Leipzig disputation of Luther, Carlstadt, and Eck in 1519 is a model of discretion in its equal praise of the learning of all participants and its careful avoidance of endorsing any one point of view.[54] A remark in the diary of his travels in France and Spain in 1526 about the bishop of Meaux's treatment of Lutherans might imply sympathy for the latter, but the statement is ambiguous.[55] In his account of his visit to Gianfrancesco Pico during his (Lange's) student years, Lange represented himself as already hostile to monks and an admirer of Melanchthon in the early 1520s, but this account was written with hindsight.[56] Yet by the mid-1540s, Lange was clearly committed to Lutheranism. In 1546–47 his long-term patron, Friedrich II, by then Elector Palatine, made Lutheran worship the religion of his territories.[57] In March 1547 Lange completed *Medicum de republica symposium*, which is dedicated to an official of the Palatinate court and contains a wide-ranging attack on Catholic beliefs and practices.[58] The *Symposium* is set up as a discussion among a theologian, a jurist, a physician, and a grammar school teacher about the merits and current problems of their respective professions. In it Lange argued simultaneously that many minor Catholic rituals (involving holy water, salt, and herbs) resembled the activities of the unlearned medical practitioners whom he deplored and that these rituals were adaptations of pagan magic. Similarly, he denounced the invocation of saints reputed to assist in particular illnesses both as ineffective and as a reversion to pagan polytheism. In short, "your whole religion, fouled with the cult of idolatry, smells of medicine carried out for money."[59]

But Lange's attitude toward magic was more complex than remarks such as these would suggest. Repeatedly, and in strongly pejorative terms, he accused medical practitioners outside the university world of using harmful magic; but he was also at pains to insist on the reality of natural magic and of sympathies and antipathies in nature, as well as to distinguish these from evil, or demonic, magic. Moreover, in Lange's view, the persistence of evil magic, even in Christian society, was at root the work of the devil, up to his usual tricks; but it was also the result of historical developments. He maintained that in Egypt and elsewhere in remote antiquity, magi were not plebeians but men of royal stock, educated in secret aspects of philosophy and the arcana of mathematics, who understood wonderful sympathies and antipathies of natural things—animals, plants, stones, and stars—and thus were able to perform works that seemed incredible to ordinary people. But the art became full of deception and diabolical tricks and was later forbidden by both Roman authorities and Christian teaching. Lange thought that in his own time the only people who really knew magic were lost souls who had committed themselves to the devil and to all the species of *goetia*, especially necromancy. Since he defined necromancy as involving, inter alia, summons to ghosts, he cited famous supposed encounters with spirits of the dead, among them both Ulysses's meeting with Achilles and Saul's visit to the Witch of Endor to raise the prophet Samuel, only to insist in each case that these manifestations were not really the spirits of the dead but instead demonic delusions.[60] Nonetheless, he was confident that in antiquity experts in good natural magic had existed: among them were Moses and Solomon, whose proficiency in magic was attested by Josephus and who surely could not have been involved in commerce with demons; rather, they were skilled in theurgy, by which they invoked the help of angels and good spirits.[61]

When it came to the contemporary world, Lange's decision as to whether to attribute any particular claimed phenomenon either to an occult natural property or to demonic intervention or (evil) magic, or simply to regard it as a false superstition, might depend on a variety of social, professional, and religious factors. Thus when Lange was ordering the pharmacist to prepare medicines made with gold and gems for Ottheinrich to use at the baths in Baden, Guillaume Rascalon—the Heidelberg student much esteemed by Lange—questioned the value of such ingredients, on the grounds that they were not used by the ancient Greeks and might be merely ostentation. As the remarks of Luigi Mondella show, Rascalon was not the only critic of the long-standing practice of using more expensive or precious ingredients in medicaments for wealthy, socially elevated patients, a type of prescription that Ottheinrich no doubt expected.[62] In response, Lange labored to distinguish between beliefs he characterized as "old woman's craziness" (*anilia deliramenta*)—for example, that wearing agate ornaments promoted marital harmony—and faculties of gems or minerals "found by experience and approved by experts" as useful in curing diseases, among them Galen's assertion that an emerald amulet strengthens the stomach.[63] In this instance, what removed a belief or practice from the realm of superstition was the testimony of an ancient medical authority. Similarly, Lange was profoundly committed to the reality of demonic intervention in human affairs. Yet he also asserted that phenomena "which the crowd attribute to witchcraft and the tricks of demons" often occurred through the power of nature—an idea that perhaps he heard at and took from Pomponazzi's lectures at Bologna. Lange's examples of these natural phenomena included concretions within tumors in the human body, since these resembled the foreign objects often said to be vomited by the demonically possessed. Yet in the same passage he went on to give a long list of what he claimed were indeed genuine cases of demonic possession and, in several other letters, reiterated his conviction of the presence and power of malign spirits. On the subject of demons, Lange seems to have only incompletely absorbed Pomponazzi's views.[64]

Above all, Lange regarded magical, or merely superstitious and ineffectual, medicine as the province of four particular categories of practitioners.

His attacks on surgeons, mainly directed at their ignorance of Greek medicine, their ineffectiveness, and their superstition, and his desire for revisions
in surgical education have been well studied by Vivian Nutton.[65] The other
three classes of individuals, named over and over again in Lange's 1545 proposal for university reform, in his *Symposium*, and in both of his collections
of *epistolae medicinales*, were monks, wandering empirics, and Jews (*monachi, agyrtae/circulatores, Iudaei*). In many respects he denounced all three
of these groups together, and in identical terms: all were *pseudomedici*, not
true physicians; there were too many of them; they knew nothing of natural causes, being totally ignorant of Aristotelian natural philosophy and
Hippocratic-Galenic medicine; they were given to magic and superstition
and made homicidal errors; and blame rested on the magistrates who allowed them to practice. Lange occasionally added women to the list, referring, for example, to "incantatrices," but by no means with the same con-
sistency as the other groups.[66] But he was clearly especially hostile to the
practice of medicine by Jews. His first medical publication, the *Symposium*,
opens with a self-authored poem denouncing Jewish physicians on religious
grounds; in the body of the work, the section on the condition of medicine
gives prominent place to the complaint that Jewish practitioners provide
medical attention for "princes, bishops, and nobles with much ostentation
and financial reward." The treatise concludes with an exhortation to the
representatives of theology and law to, respectively, threaten divine and
enforce human punishment both of *pseudomedici* and of civic authorities
who allow them to practice.[67]

Moreover, one of the longest letters in Lange's first collection of *epistolae*
(1554) expatiates on the subject of ritual murder accusations against Jews.[68]
In it he describes some of his travels undertaken apparently in the early
1540s. When he and his traveling companion, young Geraldus, arrived in
Trent, they asked their host, an old friend of Lange's, about the story of
Simon of Trent—a child whose disappearance in 1475 was blamed on the
Jews, resulting in numerous ritual murder trials, false confessions obtained
by torture, and executions; widely disseminated accounts of these trials no
doubt served to reinforce superstitions about, and hostility to, Jews. Their
host then asked Lange to expatiate on "any arcana they [the Jews] may have
both in their ceremonies and their medicaments," on the grounds that
Lange had "long been familiarly conversant with these things."[69] The re-

quest inspired Lange to reminisce about a visit he claimed to have made twenty years earlier to Gianfrancesco Pico. According to Lange's account, when the younger Pico heard that Lange and his companion were Germans, he immediately inquired about the humanist and Hebrew scholar Johannes Reuchlin, "the glory of Germany," and expressed great regret at the news of his death (which occurred in 1522) and at the consequent loss to Christian Hebraism. Lange rejoined that indeed everyone "except the monks, who have nothing in common with letters and the Muses," deplored Reuchlin's passing, and he went on to ask Pico what were "the most important books of the Jews about medicine and more secret philosophy."[70] The lengthy discourse attributed to Pico that follows sketches an account of the cabala, asserting that it originally contained "ineffable things about divinity and wisdom from the angelic intelligences" but was subsequently corrupted by "a thousand things of the art of magic . . . learned in Egypt . . . and vanities of *goetia*." As a result, no people were more deeply involved in magic than the Jews—the pious magic of Moses and Solomon having been replaced by the demonic variety, which in turn led to human sacrifice. A review of various accounts of human sacrifice from classical antiquity then leads into stories of alleged Jewish ritual murders set in the recent past, including the Trent ritual murder trials of 1475.[71]

The first part of this account, praising the wisdom of the original cabala, recalls the description in a famous oration by Giovanni Pico della Mirandola (the uncle of Gianfrancesco Pico). The idea that there were both good (natural or angelic) and bad (demonic) forms of magic, or that an originally pure magic had subsequently been corrupted, was of course by no means peculiar to Lange. Such influential recent or contemporary authors as Marsilio Ficino, Giovanni Pico, Trithemius, and Heinrich Cornelius Agrippa all espoused one or both of these ideas in some form; but their tone was not as virulent, or as pointed at a specific target, as the second part of the account that Lange put into Gianfrancesco Pico's mouth.[72] In Lange's version, the younger Pico, while admitting that some people regarded allegations of ritual murder as fabricated, insisted on the veracity of such allegations and that Jewish cabala and magic taught the cult of demons and necromancy. Lange's response (in a remark he assigned to his traveling companion) was to claim—erroneously—that this view could also be ascribed to Reuchlin.[73]

The extent to which any of this represents anything Gianfrancesco Pico actually said (even supposing the historical reality of Lange's visit to him) remains a matter for speculation. Nevertheless, Gianfrancesco Pico's views in many respects differed from those of his uncle Giovanni Pico, the most celebrated Christian exponent of cabala as a positive tradition. As is well known, the younger Pico drew on ancient skeptical tradition, as represented by Sextus Empiricus, as a means to support Christian faith; his *Examen vanitatis doctrinae gentium et veritatis Christianae disciplinae* (first printed in 1520) casts doubt on human arts and sciences and on pagan— especially Aristotelian—philosophy as ways of achieving knowledge, leaving Christian religion as the only path to truth.[74] Gianfrancesco Pico also became convinced of the reality of magic and witchcraft (and persecuted witches on his domains), strongly denounced all forms of magic, and further claimed that Giovanni Pico, too, had rejected all occult arts by the end of his life.[75] Moreover, Gianfrancesco Pico devoted a chapter in one of his works to a specific attack on Jewish magic. But that chapter is focused on illiterate practitioners who appeal to supposed books attributed to Enoch and Solomon; in it the younger Pico made no mention of either cabala or ritual murder accusations.[76]

After this reminiscence regarding the younger Pico, the rest of Lange's letter deplores the prevalence of Jewish medical practice in Germany. Lange asserted that neither Pico (presumably here the elder Pico) nor Reuchlin, "leaders in the Hebrew language" (*Hebraicae linguae antesignani*), had found that Jews had any books of medicine unknown to Christians. St. Luke's medical knowledge, attested in the New Testament, doubtless came from Greek sources available in major cities of the Holy Land at the beginning of the Christian era; but alongside this good medicine was a more sinister, magical medicine, much of which consisted of knowledge Joseph had brought back with him from Egypt. Subsequently, Lange claimed, the diaspora and troubles of the Jews would have prevented them from developing a tradition of medical learning. Lange's sweeping denial of merit, antiquity, and originality in Jewish medicine is somewhat reminiscent of Paracelsus, some of whose works were available at Heidelberg as a result of the alchemical enthusiasm of Elector Ottheinrich.[77]

The letter containing reminiscences of a visit to Gianfrancesco Pico thus situates Lange in relation both to widespread popular beliefs about Jewish magic and to leading scholars of Christian Hebraism. The perception of Jews as engaged in magic, very evident in Lange's writings, was deeply ingrained in late medieval and early modern European culture.[78] Lange's interest in Christian Hebraism as an intellectual movement (which, in any case, seldom involved favorable attitudes toward contemporary Jews) was, at most, peripheral. He was certainly familiar with some of the writings of both the elder Pico and Reuchlin.[79] But nothing suggests that Lange himself attempted to learn Hebrew or ever questioned Jewish or convert informants. Nor do his writings reveal that he had read any of the accounts available in his lifetime of Jewish ceremonies or beliefs that were produced by converts from Judaism, of which the most important was probably Antonius Margaritha's *Der gantz Jüdisch Glaub*, first published in 1530. These and later works of the same kind have been described as ethnographic, though, as modern scholarship has pointed out, they are also strongly polemical and anti-Jewish.[80] Rather, it seems highly likely that Lange had read one of Luther's anti-Jewish polemics of 1543, which in turn is said to have been heavily influenced by Luther's own reading of Margaritha.[81]

It is more difficult to place Lange's diatribes against Jewish medical practitioners in the context of his own experience of medical practice. Did his complaints about their numbers, Christian clientele, and, in some instances, influential patrons bear any relation to actual professional competition that he encountered in the Palatinate, or did these remarks simply echo already traditional polemics?[82] In the many political units of sixteenth-century Germany, patterns of Jewish settlement varied widely from place to place and time to time, with many fluctuations caused by local or regional expulsions. But increasingly, as the sixteenth century wore on, the will of the territorial prince determined policy in this as in other areas.[83] In the Palatinate, Ludwig V and Friedrich II adopted a relatively tolerant position toward Jews, at any rate as contrasted with some fifteenth-century predecessors; in 1550 there were 148 licensed Jewish families living in 88 different places, of which Heidelberg was the only major town.[84] Moreover, notwithstanding negative stereotypes (Lange was certainly not alone in his views), both Ludwig V and Friedrich II valued Jewish physicians. Substantial numbers of items in Ludwig V's vast personal collection of medical recipes were attributed to Jewish practitioners, and both rulers licensed the practice of some Jewish doctors.[85] But Ottheinrich, by the time he succeeded as Elector Palatine in 1556, had become intensely hostile toward Jews and favored their expulsion.[86] It was in this changed climate that Lange's old proposal for restricting medical practice to university graduates and to practitioners approved by the Heidelberg medical faculty became part of Ottheinrich's university reform of 1558.

Lange's collections of *epistolae medicinales* reflect his formation as a medical humanist in the first quarter of the sixteenth century, just as it seems likely that he took the idea of publishing such collections from Manardo and other Italian predecessors of that period. The epistolary form—providing for brief treatment of a diversity of subjects, displaying social interaction, and at times allowing for a lighter approach—appealed to the humanist taste for miscellanies. At the same time, in a medical context letters provided yet another format in which to satisfy the growing interest in particulars, evident in the sixteenth-century development and multiplication of such genres of medical writing as case histories, *consultationes*, *observationes*, and autopsy reports. Lange's engaging enthusiasm for Greek medicine, for natural history, for antiquities, and for accounts of travel and discovery finds numerous parallels in the medical humanists among his contemporaries. But unlike many of them—notwithstanding Lange's repeated insistence that medicine required knowledge of philosophy and analysis of causes— he seems to have had little interest in expositions on medical theory or in close readings of ancient texts. Above all, his letters, most of them probably written in the middle years of the century, reflect his experience as a practitioner and a court physician. In some aspects Lange still appears as a medical reformer: for example, in his views on uroscopy, astrology, and the need for improved surgical education. In other respects, his ideas about the profession of medicine and much else were, unhappily, not shaped by ideals of humanist learning, but instead strongly influenced by social and religious attitudes of his time and place.

HORAT. AVGENIVS. THEOL. PHILOSOP. ET. MEDICVS. PRÆSTANTISSIMVS. ANNO. SVÆ. ÆTATIS. LXIIII.

Est hæc certe hominis facies mortalis at ipsum
Diuinum ingenium est famæ imortalis et æui

M.

Columbi.

Orazio Augenio (1527–1603), at age sixty-four. This image of Augenio, reproduced here from a print in the collection of the National Library of Medicine, Bethesda, was also used as the portrait frontispiece of Augenio's *Epistolarum medicinalium tomi tertii libri duodecim: in quibus non solum maximae difficultates ad medicinam et philosophiam pertinentes dilucidantur: sed etiam Alexandri Massariae Vicentini Additamentum apologeticum et disputationes secundum Hippocratis et Galeni doctrinam funditus evertuntur* (Frankfurt: apud heredes Andreae Wecheli, Claudium Marnium, & Ioannem Aubrium, 1600). Courtesy of the National Library of Medicine.

CHAPTER THREE

The Medical Networks
of Orazio Augenio

THE MEDICAL LETTERS of Johann Lange were the product of an early to mid-sixteenth-century career in which a relatively brief period of university study and teaching led to lifetime tenure at a German princely court. Lange's intellectual formation was rooted in the first generation of medical humanism in German universities, but his court-centered professional life and his opinions were further shaped both by religious upheaval and in reaction against the regional vitality of traditions of medical practice that lay outside the world of Latinate academic medicine. By contrast, the *epistolae medicinales* written and assembled for publication by Orazio Augenio come from a very different regional and cultural context. Augenio wrote in Italy during the second half of the century and thus belonged to a medical world characterized by humanism, already long established in medicine as in the broader learned and literary culture; by some famous innovations and innovators in academic medical teaching; and by the presence of graduate physicians, even in small towns.[1] Moreover, Augenio's career trajectory also stands in sharp contrast to that of Lange, since Augenio spent almost three decades as a town medical practitioner before holding university positions for the last part of his life. As a result, Augenio's *epistolae medicinales* nicely illustrate some ways in which letter writing—and the publication of collections of letters—could play a role in building a sixteenth-century medical reputation and career in a highly competitive milieu.

Orazio Augenio (ca. 1527–1603) came from a medical family in Monte Santo, a small town in Le Marche, a region then part of the Papal States.[2] It is not known where he obtained his medical degree, but he seems to have begun both his studies in arts and his practice of medicine locally: it is claimed that he briefly taught logic at the University of Macerata, and he remembered with gratitude an older physician who guided him when he was a young practitioner at Camerino.[3] Augenio also seems to have embarked on the sort of academic itinerary not uncommon then, and to have spent some time in medical study at each of the universities of Pisa, Padua, and Rome, since he named professors at all three as his teachers.[4] Following a brief stint teaching medical theory at the University of Rome sometime from the late 1550s to early 1560s, Augenio was for many years—twenty-eight, by his own account—a medical practitioner in a succession of the small towns in his native region. He also published two of his first short printed medical works there.[5] Finally, in 1578, when already well advanced in age, he received a call to a professorship of practical medicine at Turin, a university only recently refounded by Duke Emanuele Filiberto of Savoy as he recovered his dominions from French control.[6] In 1592 Augenio succeeded in moving from Turin to the position of first professor of medical theory at the University of Padua, at that time the most celebrated center of medical teaching in Europe, where he remained until his death. Throughout the changes of these forty years, Augenio wrote and collected his "letters and consultations," or what he chose to designate as such. The first edition of his letters, published in Turin in 1579, occupied a quarto-sized volume; subsequent editions appeared as substantial folios. In their final form, Augenio's published letters amounted to three volumes, each made up of twelve internal books (for a total of almost a thousand pages in large folio); in whole or in part, they went into six editions.[7]

Although Augenio was also the author of a treatise on bloodletting that went into several editions, as well as other publications on plague, pregnancy and childbirth, and fevers, the *epistolae medicinales* evidently represent his largest investment of time and energy on works destined for print.[8] Surviving records from his university teaching appear to consist of a couple of manuscript collections of his lectures, and perhaps a treatise on the properties of wine (included in his printed *Epistolae*) addressed to his pupils at Turin, which may represent some of his instruction at that university.[9] According to Augenio's own admission, work on his letter collections caused him to delay completing other projects. In 1598, six years after he came to Padua, he acknowledged in the dedication of a third volume of his *epistolae medicinales* to the Venetian senator then serving as moderator (that is, supervisor) of the university that "many people" were expecting him to publish his work on fevers and his commentaries on Galen, both written some twenty-five years previously. For "many people" one might no doubt read the Paduan medical faculty and the Venetian authorities overseeing the university. Augenio went on to explain, indignantly and at length, why he had instead given priority to publishing the letters containing his side of an ongoing controversy with his colleague Alessandro Massaria, asserting both that he felt entitled to defend himself against Massaria's impugnations and that Massaria's interpretation of Galen was so erroneous that the public good demanded its correction.[10] Twenty years earlier, he had given a different, and perhaps equally valid, excuse for delaying publication of his work on fevers, claiming that he was awaiting the appearance of a definitive work on the same subject by an admired older colleague. Augenio's son Ilario finally sent the treatise on fevers to press after his father's death.[11] In 1573 Augenio had described "the first part of my commentaries on the books of Galen" as almost ready for publication.[12] In 1579, in the dedication of his first volume of medical letters, he promised the dedicatee, Prince Carlo Emanuele of Savoy (shortly to succeed his father as duke), that both the book on fevers and one on the opinions of Galen and Averroës were soon to appear.[13]

In general terms, Augenio's attachment to the format of "epistolae et consultationes" doubtless showed his appreciation of a genre that—like other forms of Renaissance miscellany—provided an opportunity for treating a diversity of topics and expressing personal views. But he was also evidently an attentive reader of collections of medical letters by other authors. The earliest letter among his *Epistolae*, dated 1558, records his disagreement with one of Manardo's letters.[14] He also knew Lange's letters well: he reproduced some of Lange's anecdotes, citing their source; noted that his own father and Lange had studied under the same professor at Bologna; and carefully compared the variant versions of that professor's special eye ointment possessed by the elder Augenio and recorded by Lange.[15]

In Augenio's hands the concept of letter collection encompassed individual items that were very diverse in character. Like other early modern printed collections of *epistolae medicinales*, Augenio's letters include many essays on general medical topics addressed to colleagues or disciples; some traditional *consilia*, or advice for individual patients (usually written at the request of the patient's doctor); and a strong dose of controversy, in Augenio's case often expressed in extremely acrimonious personal terms, in missives to or about opponents. But the letters also include items ranging from correspondence with family members—his father, brother, cousin, nephew, son-in-law, and son-in-law's father, all physicians except for the brother, who was a jurist[16]—to formal forensic or other reports solicited by civic or ecclesiastical authorities. Other so-called letters are, in reality, substantial short treatises divided into multiple chapters: among them, for example, are works "on the art of consultation" and on the concept of innovation in medicine, respectively nineteen and seventeen chapters in length.[17] Moreover, Augenio seems to have considered correspondence useful as a substitute for—or improvement on—traditional oral academic disputation. As he wrote to Pietro Crispo, professor of medical theory at La Sapienza, probably with reference to an oral disputation that had taken place at Rome at a time when Crispo and Augenio were either fellow students or junior colleagues: "The content of oral disputations is not very stable: for many people say they are either not clearly heard, or not clearly stated, or the speech was hurried through. But writing always remains certain. For this reason, I leave aside what you said in disputing with me, or to speak more accurately, what you wished to say, and in this letter I am writing what I think, so that if my ideas do not please you, you can boldly do what you promised you would and dispute with me."[18]

Augenio's letters reflect his medical views and interests, his friendships and enmities, and the controversies in which he engaged as they developed over time in the course of his career. They also reveal networks of contacts that he evidently cultivated carefully: with fellow practitioners in the towns of Le Marche; with the world of Roman medicine; with secular and ecclesiastical authorities; and with the Paduan academic community. These networks serve as an organizing principle for considering Augenio's rise to medical prominence, his stance toward contemporary developments in medicine and related fields, and his own claims to be a medical innovator and their limitations.

Practitioner in Le Marche

Augenio published the first volume of his *Epistolae et consultationes* in 1579, a year after his appointment to the professorship at Turin; an expanded edition with two internal volumes appeared in 1592, the year he was called to Padua. Both volumes 1 and 2 contain numerous letters written before Augenio's move to Turin and addressed to fellow medical practitioners—most of them otherwise obscure—in the towns of Le Marche or neighboring regions. Many of these letters are, or purport to be, responses to requests for a medical opinion or advice and, as such, might often circulate in manuscript beyond the original recipient.[19] Setting aside letters to family members, recipients are addressed in terms ranging from the respectfully deferential to the quasi-parental. Thus, in sending a brief treatise on the use of cautery and mustard plasters to Francesco Cirocco, *medicus praestantissimus* of Foligno, Augenio wrote: "it pleased me to address it to you, not in the expectation that you yourself would learn anything from it, since you have already reached the summit of the whole art [of medicine]."[20]

By contrast, Augenio's eight letters to Giovanni Francesco Santi or Sanzio of Cingoli are those of a mentor to a favorite disciple. Augenio was in practice in the small mountain town of Cingoli from 1570 to 1574, but his interest in Sanzio's career continued long thereafter.[21] In a series of letters written over twelve years or more, Augenio congratulated Sanzio, whom he termed *discipulus amantissimus*, on his appointment in 1570 as *medico a contratto* in a rural area, while predicting that his fame would spread throughout the region (no evidence has yet surfaced for any realization of this hopeful prediction); felicitated him on the birth of his son; assured him that he loved him like a son and missed him; and sent advice on the treatment of several patients. In a letter to another colleague he provided a long response to arguments about fevers proposed by "my Sanzio," "the most ingenious Sanzio."[22] The letters to Sanzio also offered lectures on medical topics, a mild reproof when Sanzio ventured to suggest that one of Augenio's arguments was incomplete, and general advice from the standpoint of greater experience in medical practice. Above all, Augenio urged Sanzio to be cautious in the administration of purgatives, pointing out that excessive purging could be dangerous, and not for the patient alone: "I warn that you should absolutely abstain [from extreme purgation of patients], especially at your young age, for if anything unhoped for happened to you, you would not escape from accusation of the crime of murder among the common people. It is necessary to be very cautious in practicing medicine, and especially while young."[23]

The medical content of the letters written by Augenio during his years as an urban practitioner suggest characteristics and interests that would persist throughout his career. One, as the letter just quoted exemplifies, was a fairly cautious approach to therapy, with much emphasis on danger. Another was a favorable attitude toward innovation. No significant innovations in medicine are associated with Augenio's name, but as Iain Lonie once remarked, "what is interesting about Augenio is not that he was original but that he eagerly claimed to be."[24] In the 1570s Augenio was an attentive reader of the influential French physician Jean Fernel, whom he termed "unquestionably prince among the *recentiores*"—even though his frequent citations of Fernel also include many instances of disagreement.[25] Caution and a favorable attitude toward innovation come rather uncomfortably together in Augenio's assurance—in a letter evidently reflecting a different frame of mind from that displayed in the letter quoted in the previous paragraph—that Galen's warning of the danger of purgatives for patients with diarrhea could safely be ignored, because of the superiority of modern purgatives to the remedies available to the Greeks.[26]

As one might expect, the letters Augenio wrote to fellow urban medical practitioners in the years between his departure from Rome and his arrival at Turin are overwhelmingly concerned with pathology and therapy. As usual for medicine of the period, his discussions of therapy were chiefly focused on the concept of purgation, whether by venesection or medicaments. The conditions he wrote about most frequently were kidney and bladder problems, especially calculi and urinary sediment, and fevers. He also quite often responded to questions from other physicians about women's reproductive health—that is, difficulties of pregnancy, childbirth, and menstruation or its absence. Yet despite the origin of these letters in the world of medical practice, the frequency with which Augenio discussed particular conditions is probably a better guide to which issues and controversies he thought important than to the contemporary disease environment. Throughout his career, Augenio used his letters to defend, justify, repeat, and expand upon opinions he had expressed in other contexts: in the treatises he had published or would publish on bloodletting, fevers, and plague; in oral disputation or arguments; and in other works that he claimed to have written.[27] In particular, he described his first volume of letters as published "against an importunate cynic," a play on the name of

Giulio Cini of Colle. Augenio's quarrel with Cini seems to have begun in 1572, when Cini and another physician, Vicenzo Cibo, accused one another of causing the death of a patient and Augenio vigorously defended Cibo. Matters were exacerbated a few years later when Cini attacked Augenio's views on kidney stones and ulcers and when a patient, the apostolic secretary Egidio Franceschini, ignored Augenio's recommendations in favor of the treatment suggested by Cini. In the mid-1570s, Cini and Augenio traded indignant treatises on the case and on the general topic of the treatment of kidney and bladder complaints. Augenio included in his contribution a copy of Cini's *consilium* for Franceschini, to which he added marginal numbers indicating, in his view, Cini's forty-five errors regarding the nature, causes, symptoms, and treatment of ulceration of the bladder; he also composed a "Historia propositi affectus" in ten chapters, the last chapter of which explains and corrects each of those errors. Many of the letters in the first volume of Augenio's collection involve denunciations of Cini, thanks for others' support against him, or efforts to recruit such support.[28]

The assumptions and methodology of this group of Augenio's letters are those of Renaissance Galenism (without necessarily repudiating all aspects of medicine's medieval past). Augenio and his correspondents were profoundly concerned with the analysis, application, and sometimes modification of Galenic teaching in medical practice. But other topics and areas of knowledge cultivated in contemporary academic centers of medical learning are notably absent from the letters Augenio sent to correspondents within his regional network of fellow urban practitioners. Even general expositions of broad aspects of Galenic physiological theory (for example, the functions and interrelation of humors, elements, and qualities) are relatively rare. On a very few occasions Augenio discussed natural history or antiquities, but he did so exclusively in connection with particular remedies. Thus, in an attempt to re-create authentic versions of *ptisana* (a decoction of grains) and *mulsa* (honeyed wine), medicinal drinks discussed by Hippocrates and other ancient authors, Augenio endeavored to identify the grains used by the ancients for *ptisana* and to suggest modern equivalents, and he discussed at length ancient and modern recipes for the preparation of both beverages.[29] But the letters in this group have almost nothing to say about other areas of contemporary medical knowledge or interest less directly related to practice—for example, philological criticism of Greek medical texts, the intersection of medicine and natural philosophy, or anatomy. Augenio's engagement with such topics and concerns emerges almost exclusively in letters to correspondents in his other networks. Even then, his attention to philological critique was minimal; in the three volumes of his letters I have noticed only one instance, an objection to Linacre's rendering of a sentence of Galen.[30]

Maintaining Roman Connections

The connections of the Augenio family with Rome and the world of Roman medicine began in the time of Orazio's father Lodovico, who had been one of the physicians of Pope Clement VII (reigned 1523–34).[31] Orazio revered Lodovico as his own teacher and many times drew attention to their kinship and to Lodovico's medical wisdom. Orazio's own brief teaching career at Rome was a source of pride that he did not wish to be forgotten; when the Paduan professor of philosophy Arcangelo Mercenario referred to Augenio as "new to declaration (*novus ad dicendum*)" (probably around the time the latter got his Turin appointment), Augenio responded tartly: "a man who long before taught publicly at Rome is not 'new to declaration.' "[32] Orazio's attachment to Rome also emerges from a letter to his brother Fabrizio, a doctor of civil and canon law who had remained in the city in the service of the church after Orazio left. In 1574, when Orazio heard the news of Fabrizio's appointment as vicar general to the bishop of a southern diocese, his response was to lament his brother's departure from Rome, while simultaneously extolling the health benefits of drinking Tiber water and adding a reminder of the story that Pope Paul III never traveled without a supply of it.[33]

It seems likely that regional as well as family connections helped Augenio both to enter the world of Roman medicine in the first place and to maintain his contacts in Rome after he left his position in the city. Several of the prominent physicians active in Rome whom he praised most highly or with whom he corresponded long after he permanently left Rome came from his own region of Le Marche in the Papal States: among them were the anatomist Bartolomeo Eustachi; Andrea Bacci, author of well-known works on waters and springs and of a treatise on the unicorn; and Antonio

Porto, one of the participants in the autopsy of S. Filippo Neri.[34] Another physician in Rome to whom Augenio was close was Girolamo Cordella (d. 1595), the addressee of Augenio's hotly contested advice on kidney and bladder conditions and his supporter against Cini in the ensuing controversy. Cordella—whom Augenio described as "a most perfect medical doctor, in whose teaching, diligence, and observation I have full confidence" and mentioned with respect in several letters—was physician first to Cardinal Alessandro Farnese and then to Pope Clement VIII.[35]

But Augenio also wrote to other leading Roman physicians. He carried on epistolary disputations with Pietro Crispo about the interpretation of Hippocrates's teaching on pleurisy and about the concept and proper practice of revulsion (carrying out bleeding or some other form of purgation from a site distant from the affected area of the body). To Alessandro Petroni, who was attending physician—and who taught students—at Rome's Ospedale di Santo Spirito, Augenio wrote in very different terms. At the request of an unnamed cardinal, he sent Petroni an opinion on the case of a woman patient in which he explained that in his, Augenio's, view every detail of the diagnosis and treatment recommended by Petroni and his colleagues was completely misguided.[36]

Moreover, Augenio took care to preserve and refresh his contacts with Roman medicine by making a number of return visits to the city (one of them for the jubilee of 1575).[37] Remarks in several of his letters announce—indeed, insist on—his familiarity with the most striking and innovative aspects of medicine in Rome: leading physicians, current developments in anatomy, and notable therapeutic successes. He emphasized the progress of anatomy at Rome, recalling a difference of opinion about a point of anatomy that arose "when I was professing medicine at Rome" between "the most outstanding anatomists" Eustachi and Realdo Colombo and citing dissections performed by Colombo at the hospital of S. Maria della Consolazione in an endeavor to resolve the issue.[38] Thus in 1576, writing to a surgeon on the treatment of bladder calculi, Augenio dwelt on the dangers of surgery and the many instances of death or permanent injury caused by attempts to "cut for stone" undertaken by practitioners with an insufficient knowledge of anatomy. But he added that at Rome he had seen not only a rare instance of successful treatment of this condition with a remedy taken by mouth (the patient was the young son of a papal printer) but also many patients whom the anatomist Costanzo Varolio had successfully treated by surgery. In a letter written several years later from Turin to Sigismund Kolreuter, physician of the Duke of Saxony (one of Augenio's very few letters to a correspondent outside Italy), Augenio once again described the successful medication of the youth with calculi, this time adding more detail—including the claim that he himself had participated with Juvenal Ancina (a physician who became an Oratorian and a companion of S. Filippo Neri) in treating the case—while again emphasizing the extreme rarity of such a cure. In the same letter Augenio recommended that, if cutting for stone had to be done, the knives should be heated, and he further claimed that, thanks to his suggestion, one of his students at Rome had successfully operated in this way. The accounts of the treatment of the printer's son must refer to one of Augenio's return visits to Rome, probably that of 1575, since Juvenal Ancina came to Rome in 1574 and Varolio is said to have arrived from Bologna in 1572.[39]

Augenio's carefully preserved contacts with and frequent references to Rome did not secure him the recall to the city that was perhaps part of their intention. But certainly they could only have served to enhance his reputation, and they may have played a part in his call to Turin.

Expertise at the Service of Secular and Ecclesiastical Authorities

In the years in which Augenio was building his reputation among fellow urban practitioners in Le Marche, ecclesiastical authorities already occasionally turned to him for an expert professional opinion. One cardinal instructed Augenio to send Alessandro Petroni advice about the treatment of one of Petroni's patients, and another request came from the vicar general of the diocese of Osimo, who asked Augenio to determine the cause of death in the case of a man who had suffered a slight head wound. In response, Augenio determined that, though it was possible to die of a slight wound, in this instance death resulted from a preexisting condition. More significantly, Cardinal Annibale Bozzuto sought a professional opinion on topics to which Augenio devoted special attention: the duration of pregnancy and fetal viability. The immediate context for the request perhaps was a dispute about the legitimacy of an infant. In his reply on this occasion, as in his separate treatise on the subject titled *De hominis partu libri duo*, Augenio argued strongly against the ancient belief that, for astrological or other reasons, a child born in the eighth month of pregnancy could never survive.[40]

Following Augenio's university appointments, first at Turin and later at Padua, official demands for his expert opinion multiplied from both ecclesiastical and secular authorities. At Turin, he provided such opinions for the Senate of Piedmont, one of the principal administrative agencies of the dukes of Savoy, as well as for other ducal officials. In several instances in which potential criminal accusations were involved, he displayed his habitual caution. Just as he had concluded that the victim of the head wound at Osimo had died of other causes, in two cases of suspected poisoning he assured, respectively, the Piedmont Senate and the prefect (the chief official in charge of criminal justice) of the city of Rome that the deaths had resulted from disease.[41] During an epidemic, when suspicion arose that oil and salt imported from Genoa had brought the contagion to Turin, the chancellor of the Duke of Savoy sought a professional opinion from Augenio. His report exonerated both substances, while pointing out that the process of obtaining these items from elsewhere might involve human contact with unwholesome air, infected persons, or fomites; at the same time, he cautiously distanced himself from rumors that the salt had been deliberately poisoned.[42]

Augenio was considerably less restrained in another report to the Piedmont Senate, in which he mounted a forceful defense of a physician who was threatened with expulsion from the Turin College of Physicians for compounding his own medicines, an activity that, in the view of the college, derogated the nobility of medicine as a learned profession. As this man was well known for a particular compound—his extract of rhubarb—it seems likely that the College of Physicians viewed him as violating the distinction between physicians, who offered learned advice, and charlatans, who peddled their own special or secret remedies.[43] But Augenio insisted on the scientific value, and dignity, of technical knowledge of medicinal ingredients and their preparation. Invoking the anxieties often expressed by humanist physicians about the dangers of relying on ignorant pharmacists to compound medicines, he pointed out that Turin, unlike "outstanding cities of Italy," so far had no law requiring the supervision of pharmacists by an officially deputed physician—a remark that may have struck some nerves at a time when Emanuele Filiberto and his successor and their advisors were doing everything in their power to magnify the status of Turin. Augenio perhaps had in mind the arrangement for the supervision of pharmacists in Rome, where the *protomedico* and the College of Physicians were responsible for licensing and supervising all medical practitioners and apothecaries in the city. He clinched his case for the defense by providing lists of famous modern physicians who compounded their own drugs, pharmacological writers who praised rhubarb's medicinal virtues, and princes who had used it and benefited thereby.[44] On another occasion, in a report to the chief physician of the Duke of Savoy, Augenio mounted an equally strong defense of a surgeon who was accused of causing the death of a patient, vigorously denying that the double cautery performed on both the patient's arms could have been responsible for his demise.[45]

In general, Augenio seems to have won the confidence of ecclesiastical as well as secular authorities as a source of reliable medical opinion. Although he addressed a handful of letters to Theodor Zwinger in Basel in the 1580s, Augenio's contacts with the Protestant republic of medical letters of northern Europe were few and slight—at any rate as compared with the contacts maintained by some of the other Italian physicians in the same period, notably the extensive correspondence between Mercuriale and Zwinger— and appear to have aroused no suspicions as to his orthodoxy.[46] A number of the patients for whom he prepared written *consilia* or about whom he consulted with medical colleagues were clerics or members of religious orders, including the papal nuncio to the Duke of Savoy, another bishop, a humanist apostolic secretary, a Franciscan theologian, and several nuns. Among the latter was a sister in a convent in Genoa, twenty-three years of age, who had been coughing blood for two years, recently in large amounts, while she remained responsible for distributing bread to the entire community. Her treatment, which consisted of extensive bloodletting, gave rise to vigorous disagreements among the attendant physicians about the proper method of doing so. Augenio, unlike the Spanish physician who had requested his opinion, regarded the coughing of blood as a more serious problem than the cessation of her menses.[47]

Moreover, on at least one occasion Augenio submitted testimony as an expert witness to the Auditors of the Roman Rota, the highest ecclesiastical judicial tribunal. That case was one in which dissolution of matrimony was sought after one year of marriage, on the grounds of the husband's impotence. The couple in question were on extremely bad terms: after the young husband hit his wife, she had developed "immortal hatred" for him and gone home to her parents, vowing never to return. By the time Augenio gave his opinion, four physicians appointed by the couple's bishop had already reported it as their judgment that the husband was incapable of intercourse with his wife and thus the marriage should be dissolved. But Augenio contradicted the first set of physicians, arguing vehemently that the husband was not impotent and opposing the dissolution of the marriage. With some reason, he suggested that the husband's capacity for marital intercourse might have been temporarily—though not necessarily permanently— removed by quarrels with his wife and her parents and by intrusive inquiries into his sexual performance on the part of "the midwife, the archiepiscopal notary, and four doctors."[48] Augenio subsequently wrote up the same report into a formal treatise in ten chapters, and this, too, he incorporated into his *Epistolae medicinales*. In this second version, the story of blows and family quarrels is minimized in favor of disquisitions on male physiology and a new emphasis on the likely role of witchcraft in the husband's temporary impotence. Belief in the power of magic to cause impotence was widespread in medieval Europe and found its way into medieval learned writing on theology, canon law, and medicine; but physicians seem to have begun to discuss particular cases with more frequency in the fifteenth and sixteenth centuries, when the general climate of anxiety about witchcraft was combined with a growing role for narrative and *historia* in medicine.[49]

Yet Augenio's pronouncements on one aspect of either this or another matrimonial case did indeed bring down on him the disapproval of a judge of the same tribunal. In 1587 Augenio felt obliged to address a long, self-exculpatory treatise to Serafino Olivieri, the dean of the Roman Rota, defending views on female anatomy that Olivieri had characterized as "rash." Augenio denied that there were any certain, physical signs of virginity universally found in all women. In the treatise to Olivieri, after lining up arguments and ancient, medieval, and recent medical authorities on both sides of the question, including the anatomical dispute between Eustachi and Colombo on the same issue, Augenio rejected the identification of the hymen as an anatomical organ, characterizing it instead as a "preternatural" condition of some young women; and he was equally dismissive of the belief that bleeding invariably followed first intercourse.[50] Augenio's arguments bring to mind the similar review of divergent opinions about the hymen among anatomists and skepticism regarding the reliability of inspections for virginity by midwives found in Laurent Joubert's *Erreurs populaires au fait de la medecine et regime de santé*, first published in Bordeaux, Paris, and Avignon in 1578. Whether Augenio knew of the discussion in Joubert's *Erreurs populaires* at the time he wrote the response to Olivieri is unclear; he does not appear to cite Joubert directly, and the Italian and Latin translations of Joubert's work postdate Augenio's response to Olivieri. But a decade later, in 1598, Augenio listed Joubert among praiseworthy modern innovators, "men most distinguished in every branch of knowledge, and most modest and most prudent."[51]

Olivieri took Augenio's position as an attack on canon law, which provided for physical inspection for virginity by midwives or matrons in some disputed matrimonial cases. Augenio defended himself from the imputation of "contradicting the papal law" by asserting that he had merely meant to say that such inspections were far from infallible, that their results were at best only probable, and that canon law itself acknowledged this by providing for repeated inspections in cases of doubt.[52] In one of the few instances in which he denounced an entire group of practitioners, he turned the argument into an attack on the limitations of midwives (or, at any rate, those in contemporary Piedmont), whom he characterized as "rudes mulierculae" (ignorant little women), very different from the women expert in medicine respectfully described by ancient authors. He added the remarkable suggestion that midwives should dissect bodies, or at least have been present at dissections, before undertaking their office, adding—with how much justification I do not know—that midwives in Spain were obliged to view dissections.

Building Paduan Connections

Augenio was in contact with members of the medical faculty at the University of Padua long before he received the call to join them. In 1576 he was pleased to learn that Girolamo Capodivacca, professor of *medicina practica* at Padua, thought well of his views on calculi and kidney disease; as a result, he sent Capodivacca a treatise on the subject in the hope of enlisting him as an ally in the quarrel with Cini.[53] He credited Girolamo Mercuriale, a leading member of the Padua medical faculty from 1569 to 1587, with recommending him for the Turin position; a grateful acknowledgment in the dedication of Augenio's first volume of medical letters characterizes Mercuriale as "the most perfect physician of our time, who with outstanding praise rightly holds the first place at the University of Padua."[54]

But some other early contacts with leading members of the Padua professoriate were considerably more ambiguous. One exchange in particular is illuminating with regard to the hazards as well as the possible benefits of establishing professional contacts by correspondence and of publishing medical letters. Augenio came from Monte Santo, the same hometown as Arcangelo Mercenario, professor of philosophy at Padua and the author of numerous Aristotelian commentaries, and Augenio was always attentive to regional and local connections. In 1576 the news of Mercenario's forthcoming promotion to a senior position inspired Augenio to write enthusiastically to a friend: "I certainly not only love him but observe how much he is the glory and ornament of all the literary men of our province."[55]

Shortly before Augenio moved to Turin, he was surprised to receive a letter from Mercenario asking his opinion on "some philosophical and medical problems," problems for which Mercenario claimed that he had never found a satisfactory solution, either in oral disputations at Padua or elsewhere, or in writing. Mercenario, of course, inhabited an academic milieu in which Aristotelian natural philosophy and Galenic medical theory remained in close contact, both institutionally and intellectually. His questions concerned the correct interpretation of some of Galen's more obscure statements about primary qualities in relation to disease, remedies, cure by contraries, and the medical concept of degree: what, really, did it mean to say that remedies should be "equal"? If they were equal to the disease, how could they overcome it? How should Galen's concept of resistance among qualities—for example, the statement that cold has less power of action than heat, but more resistance—be understood? The letter arrived at an awkward moment for Augenio, and he postponed a substantive response until after he was settled at Turin. Then he published his reply, together with Mercenario's original inquiry, in his first volume of letters in 1579.[56] Alas, in late 1582 Augenio received a letter from Padua informing him that his response had met with denunciation in a work recently published by Mercenario. The author of this letter was Michele Colombo, whom Augenio described as "discipulus meus carissimus."[57] Because Colombo was also one of Mercuriale's students and editors, Augenio's description of him suggests another early link with Padua, and, in addition, Augenio's good relations, during this period of his career, with Mercuriale.

On hearing of Mercenario's rejection, Augenio this time responded immediately, with a lengthy "Apologeticus against Arcangelo Mercenario," which he both appended to a new and enlarged edition, published in Turin, of his work on bloodletting and included in the second volume of his letter collection.[58] The "Apologeticus" contains a greatly expanded exposition of the arguments in his earlier response, prefaced by a verbose explanation of the personal and family problems—his father's death, the need to clear up family affairs in Monte Santo, the logistical complications of the move to Turin, and the heavy workload when he got there—that had caused him to delay replying to Mercenario for several years.[59] But Augenio also admitted other reasons for his delay. He had professed to be both deeply flattered and greatly surprised by Mercenario's approach to him, but he also suspected—and subsequently confirmed—that Mercenario's letter was sent "as a test of my intellectual powers"; and Augenio was painfully aware that in his long years in medical practice he had seldom been called on to engage in the kind of academic medical/philosophical argument that Mercenario proposed.[60] After this epistolary encounter, perhaps Augenio sensed the need to shore up his philosophical and Aristotelian credentials. In any event, just three months after writing the "Apologeticus," he felt impelled to send Mercenario another long letter disputing the latter's views on Aristotle's theory of concoction—that is, perfection by the application of heat, a key concept in explaining physiological processes for both Renaissance Aristotelian philosophers and contemporary Galenic physicians.[61] Whether Mercenario's initial letter posing problems to Augenio was perhaps intended as a first tentative approach on the part of the University of Padua remains unclear. But the episode shows that the real benefits of participation in medical correspondence and published collections of medical letters in building professional communities and opening opportunities could also be accompanied by equally real hazards to one's scholarly reputation.

Nevertheless, in 1592 Augenio received his call to the position of first professor of medical theory at the University of Padua. By that time, of the members of the faculty who had previously paid the most attention, favorable or unfavorable, to him and his work, Capodivacca and Mercenario had died and Mercuriale had left. It seems likely that Alessandro Massaria, Mercuriale's successor in the position of first professor of practical medicine, had some part in suggesting Augenio's appointment. So, at any rate, Massaria claimed, although he and Augenio rapidly became involved in the controversies with which Augenio filled the third volume of his medical letters. It was a claim, however, that left Augenio sputtering with indignation: "In sum, Alessandro Massaria, it was not you that brought me to this school, but my own hard work and nights devoted to study!"[62] Augenio's arguments with Massaria largely revolved around Massaria's criticism of Augenio's writings on bloodletting, of which several further, enlarged editions appeared during Augenio's years at Padua.[63] The details of their differences of opinion on this topic need not be pursued here, though the quarrel continued until Massaria's death—which Augenio wrote that he regretted, because he had been looking forward to overwhelming his old enemy with a forthcoming rebuttal.[64]

Augenio's wrangling in print with Massaria over bloodletting was sufficient to dismay Mercuriale, who had been a warm supporter of Augenio at an earlier stage of his career. Commenting on the dispute in a letter (probably written from Pisa, where Mercuriale taught from 1592 to 1598) to Fabrizi ab Acquapendente, Mercuriale stated:

In these last days I have read through the discussions of Orazio Augenio, a most learned man and a very good friend of mine, on bloodletting, and [his participation] in that disputation about the indications for bloodletting that has been sufficiently eloquently treated by many people. When I was getting to the end I wondered greatly that it should so often happen that when people strive with great ingenuity to shed more light on things that are obvious in themselves they make them more obscure and difficult. This is something that outstanding men and of great authority should make every effort to avoid[;] . . . they depart from that eternal precept of Aristotle in *Rhetoric* where he teaches that men naturally choose ease and brevity of verbal expression and at the same time commands the single virtue of diction to be speaking plainly, lest anyone who undertakes to teach others be long, difficult, and obscure. And certainly (I speak freely) having very briefly examined this controversy about bloodletting, entered into by some people who are, indeed, most skilled, I can scarcely bear to be consumed by so much ingenuity and eloquence without in the end obtaining anything other than [an awareness of] a certain vast desire of fighting, and of drawing out disputations, not without some damage to the souls of the studious, especially when it would be legitimate for us to end the controversy with a few brief points if we abandoned quarrels.[65]

Mercuriale then proceeded to demonstrate what he meant about brevity by summarizing his own interpretation of the views on bloodletting of Hippocrates and Galen in the remaining three pages of the letter to Fabrizi.

But Augenio and Massaria also differed on another issue, potentially larger in its long-term implications: the value of innovation in medicine. In 1587 Massaria had begun his own tenure as professor of practical medicine at Padua with an inaugural lecture in which he proudly asserted his medical humanist credentials, proclaiming himself a follower of Galen and the Greeks, distancing himself from those who still adhered to the medieval Arabo-Latin medical tradition, and roundly denouncing followers of "none of the ancients, but . . . some more recent doctrine."[66] Augenio's answer came in 1598, when, for the third volume of his medical letters, he composed a treatise that responded to Massaria with a defense of innovation and innovators in medicine.[67] Like many academic quarrels, Augenio's response to Massaria on the subject of innovation and innovators had a personal as well as an intellectual dimension. Massaria had named two midcentury figures as the principal examples of pernicious modernism: Jean Fernel and Giovanni Argenterio, the latter best known for his sharp attacks on Galen's writings on disease. Augenio was an appreciative reader of Fernel and a warm admirer of Argenterio, who had been his teacher at Pisa and among his predecessors at Turin. Among other expressions of respect, Augenio claimed that he had delayed publication of his work on fevers because he was waiting for the appearance of Argenterio's writings on the subject, which he believed would leave nothing more for him to say.[68]

Augenio's treatise on innovators in medicine, *De medicis novatoribus disputatio*, is, for the most part, an indignant attack on Massaria, informing him at length that he was not as good a Galenist as he thought he was. The tone of Augenio's remarks is sufficiently exemplified by the following passage:

> You ask, "Why therefore do you dispute with me?" Because, I say, to connive and assent [to error] when the matter concerns human life is impious and detestable. Who could bear it calmly were he to see the institutions of Hippocrates and all the ancients subverted? When you openly corrupt Galen's teaching of which you claim to be a follower to the least detail? When you condemn and detest the common opinion of all the Arabs and all the Latins? When you declare that all those who have up to now written about indications were wickedly deceived and asleep with their eyes tightly closed? When you persuade yourself that it is all right to mutilate Galen's words everywhere? Even whole sentences? Even citing the context wrongly? When you think so badly of all the professors of this university? When you publicly proclaim them to be tiny apes of men of letters? When you want to be the only person in this school with the reputation of a second Galen?[69]

But Augenio's treatise also incorporates a qualified defense of the new, and of change, in the art and science of medicine. In its openness to innovation this work may go somewhat further than Augenio's proposal, made a few years later, for a new course in the Padua medical curriculum on diseases, symptoms, and diagnosis. In that instance, too, long experience of medical practice combined with receptiveness to some critics and critiques of Galen may have provided the inspiration; but the actual text Augenio suggested for the proposed course was a section of the medieval *Canon* of Avicenna.[70] In the treatise on innovators included among his medical letters, Augenio proclaimed that "in my opinion this name of 'innovator' ought not to be considered hated and detestable" and that innovators were not, as Massaria alleged, "corrupters of the republic of letters." He divided innovators into those who challenged only some sections of Galen's doctrine and those who proposed radical change. In the first category, he placed Avicenna and Averroës and a list of sixteenth-century writers—prominently including, of course, Fernel and Argenterio.[71] Discussing the ideas of recent critics of some parts of Galen's teaching could only be beneficial, as even if their claims were false, disputation against them might uncover the truth (Augenio seems to have had only disputation in mind, not any form of experimental or practical verification; his list of moderns does not include Vesalius).[72]

Augenio's second category of truly radical innovators consisted of "masters of the art (*artifices*) who, having condemned all the opinions of their predecessors, introduced a new art and science." This second category contained just three individuals: Hippocrates, Aristotle, and Paracelsus. Of Paracelsus, Augenio declared that he "founded a new sect, that is, chymical medicine, as they call it." He added that although Paracelsus had not, as some of his followers claimed, invented the art of distillation, he "deserved praise because he seemed to have called that art back into medical use in our time and greatly amplified it."[73] Although knowledge of Paracelsian remedies was diffused in late sixteenth-century Italy, especially among empiric practitioners, and interest in chemically prepared medicines was widespread, Paracelsus had very few admirers among the strongly Galenist professors of medicine in the Italian universities. Augenio's evaluation—for all its qualifications—is sufficiently positive to merit notice. Augenio has, on occasion, been classified as an anti-Paracelsian, because in his treatise on consultation he stated that a Galenist should not consult with a Paracelsian, as they proceeded from completely different principles. Moreover, as Richard Palmer has rightly pointed out, consulting with Paracelsians could have a negative effect on the professional standing and income of a Galenist physician, a consideration that may explain Augenio's comment in the piece on consultation.[74] But, in the light of *De medicis novatoribus disputatio*, his comment in the former treatise may also be read as simply a statement of the difference between the two systems rather than an outright condemnation of Paracelsianism. Moreover, it is perhaps worth noting that Augenio retained his positive attitude toward innovation into old age, since he was over seventy when the treatise on *novatores* was written.

Orazio Augenio belonged to a relatively small group of sixteenth-century medical men for whom the writing, collection, and publication of medical letters was their primary means of written expression. Obviously this was not the case for all authors of published collections of medical letters—for example, for Gesner and Mattioli letters were only one part, and not necessarily the most significant, of a much larger literary output. To be sure, Augenio used the published letter collection as a catchall format in which it was convenient to include short treatises that might otherwise have appeared independently (if inconspicuously), as well as more standard epistolary essays and *consilia*. Letters also provided him with a means of recruiting supporters in his controversies with colleagues, controversies that, in his case, always seem to have inextricably mixed a good deal of personal animus into disputations over issues in medicine. But Augenio also gave repeated indications that his decision to focus on published letter collections was accompanied by an awareness that the consequent slender production of published work in other, more conventional, medical formats (handbooks of practice, commentaries on standard authors, treatises on diseases, and so on) might involve a level of professional risk.[75]

Yet the publication of letter collections served Augenio well. Indeed, one can read the stages of his ascending career through his letters. For example, he proudly indicated the five editions of his treatise on bloodletting.[76] But there seems little doubt that Augenio developed and built his professional reputation, and ultimately achieved an elevated academic position in the medical faculty of a major university celebrated for medical teaching, chiefly through his carefully maintained personal and correspondence networks: within a regional circle of fellow practitioners in the small towns of Le Marche and, farther afield, with academic colleagues and ecclesiastical and secular authorities in the principal centers of learning and administration in the Papal States, in Savoy Piedmont, and in the Veneto. Moreover, although it is clear that some of Augenio's letters circulated in manuscript, it was the publication in print of large collections from his correspondence that confirmed his renown. His collected letters eventually also reached audiences outside Italy—all three volumes, including the first edition of volume 3, were published in his lifetime by the house of Wechel in Frankfurt. As Ian Maclean has pointed out, Wechel and his heirs were speculative publishers who studied the market carefully; they would presumably have been unlikely to take on Augenio's *epistolae medicinales* if they did not foresee that they would find readers in northern Europe, just as they had done in Italy.[77]

Conclusion

PUBLISHED COLLECTIONS of medical letters were at least in part a humanistic genre in origin, inspired, like compilations of Renaissance letters relating to other areas of knowledge, by interest in ancient letters and letter writing, enthusiasm for the recovery of ancient literary forms, and appreciation of the openness and diversity of a format appropriate for all kinds of inquiry and critique. From this standpoint, the collections of letters published by Giovanni Manardo of Ferrara and other learned physicians in early sixteenth-century Italy may be regarded as products of medical humanism; hence the encomium to the purity of Manardo's language by his fellow citizen, the humanist Celio Calcagnini—and hence, too, Calcagnini's somewhat defensive insistence that practical usefulness outweighed the lack of literary elegance in most other medical writing.[1] As printed collections of medical letters multiplied across the sixteenth century in both southern and northern Europe, the genre clearly seems to illustrate a growing importance attached by physicians and naturalists to attributes associated with the concept of the Republic of Letters: open and relatively informal communications among a learned community, a liberal exchange of information and ideas, and, often, mutual reinforcement of professional reputations.

But letter collections also reflected widely differing types of medical experience, and their authors or editors used them for a variety of purposes. An editor who assembled a group of letters by multiple authors "gathered from the most distinguished philosophers and physicians of our age" (as the title of Scholz's collection put it), a son or pupil of a physician who collected his father's or teacher's letters for a commemorative or celebratory volume, and a physician who gathered his own letters for publication were each engaged in a different enterprise. Moreover, the transition from script to print was likely to be affected by external circumstances, whether relating to printers and expected readership (as the work of Ian Maclean has shown) or to the current political or religious environment. In the case of letters exchanged across confessional and political boundaries, no doubt the original correspondence served to keep the writers in touch with intellectual developments in other regions of Europe. Yet intensifying religious censorship increasingly required Italian medical authors to balance prudence and risk in maintaining contacts with professional colleagues in centers of reformed religion north of the Alps. After midcentury, with rare exceptions, published examples of correspondence between professors of medicine at Padua and their German colleagues appear only in collections printed in the German lands. In that context, no doubt, such letters served to illustrate close intellectual contact and Galenic medical orthodoxy shared by the northern writers and leading professors at the most famous center of medical learning.

Whatever their place in the literature of medical humanism and in the Republic of Letters, sixteenth-century medical letter collections diverge widely in style and content. This is so on account not only of the personal interests and life situation of each author or compiler, but also of the varied extent to which the different printed letter collections resemble or assimilate older genres of medical writing, most notably that of collections of *consilia*, and older forms of academic debate, especially the practice of disputation. The boundaries between collections of letters, groups of *consilia*, and assemblages bearing some such title as *consultationes et responsa* are by no means sharply delimited (as the case of Girolamo Mercuriale's *Consultationes* in chapter 1 may suggest). Of the two authors examined in some detail in chapters 2 and 3, Lange's letter collections are both more personal and more infused with literary artifice than many other examples of the genre. They provide some vivid illustrations of the life, daily occupations, and social contacts of a court physician, and of the variety of demands that could be made on him. At the same time, Lange's fondness for incorporating dialogues into his so-called letters, his attention to literary and moral-philosophical as well as medical sources, and his interest in travel accounts and antiquities all link him closely to a world of humanism beyond medicine. Less concerned with disputation with individual colleagues than many other authors of *epistolae medicinales*, Lange turned his energies against entire groups of practitioners outside the charmed circle of Latin and humanistic medical learning. In the context of the official Lutheranism of the midcentury Palatinate, his religious and professional prejudices could easily—and unfortunately—coincide. Such attitudes were by no means peculiar to him, but among writers of medical letters he certainly expressed them with a notable intensity.

With Augenio's pragmatic letters, many of them reflecting his concerns as a practitioner, we may be seeing the end of the world of medical humanism. Augenio, like Lange, made very free with the concept of what constituted a letter; but Augenio's practice of incorporating entire short treatises on medical topics into his so-called letters is a far cry from Lange's humanistic dialogues. Rather than the world of the court, many of Augenio's letters portray urban medical practice, from which a long, slow climb led him to the patronage of rulers, both ecclesiastical and secular, and the world of universities. Augenio's letters eloquently illustrate both his range of strictly medical interests, including a striking appreciation for innovation in medicine and anatomy, and his propensity—shared with many other authors of *epistolae medicinales*—for turning the disputational aspect of the medical letter into a series of fierce controversies with individuals perceived as rivals or critics. And Augenio's focus, never wider than regional, was intensely local by the last part of his life, when he had attained a chair at Padua. In the 1570s and 1580s, members of the Paduan medical faculty had corresponded with physicians at the courts of Vienna and Prague and the universities of Heidelberg, Zurich, and Basel; twenty years later, Augenio concentrated his energies on collecting and publishing a volume of letters largely devoted to attacking a colleague at Padua.

Many other collections of *epistolae medicinales* await detailed investigation; indeed, much more could be said about the few examples discussed in this study. I hope, however, to have suggested some of the ways in which sixteenth-century collections of medical letters assembled for publication throw light on some major characteristics, or types, of the genre and on the variety of professional and intellectual milieux of their authors and editors, as well as on aspects of their medical knowledge and practice reflected in these collections.

Abbreviations

ADB   *Allgemeine Deutsche Biographie.* 56 vols. Historische Commission,
Königlichen Akademie der Wissenschaften, Munich. Leipzig: Duncker
& Humblot, 1875–1912. Reprint, Berlin: Duncker & Humblot,
1967–71.

DBI   *Dizionario biografico degli italiani.* Rome: Istituto della Enciclopedia
italiana, 1960–.

NDB   *Neue Deutsche Biographie.* Historische Kommission, Bayerische
Akademie der Wissenschaften, Munich. Berlin: Duncker & Humblot,
1953–.

Introduction

1. I make no attempt to provide a complete bibliography on these subjects.
The following are some of the works I have consulted on the topic of letters and
letter writing: A. Gerlo, "The *Opus de conscribendis epistolis* of Erasmus and the
Tradition of the *Ars epistolica*," in *Classical Influences on European Culture A.D.
500–1500: Proceedings of an International Conference Held at King's College, Cambridge, April 1969,* ed. R. R. Bolgar (Cambridge: Cambridge University Press,
1971), 103–14; Cecil H. Clough, "The Cult of Antiquity: Letters and Letter Collections," in *Cultural Aspects of the Italian Renaissance: Essays in Honour of Paul Oskar
Kristeller,* ed. Cecil H. Clough (Manchester: Manchester University Press, 1976),
33–67; Marc Fumaroli, "Genèse de l'épistolographie classique: Rhétorique humaniste de la lettre, de Pétrarque à Juste Lipse," *Revue d'Histoire Littéraire de la
France* 78 (1978): 886–905; Judith Rice Henderson, "Erasmus on the Art of Letter-
Writing," in *Renaissance Eloquence: Studies in the Theory and Practice of Renaissance
Rhetoric,* ed. James J. Murphy (Berkeley: University of California Press, 1983), 331–
55; eadem, "Defining the Genre of the Letter: Juan Luis Vives' *De conscribendis*

epistolis," *Renaissance and Reformation*, n.s., 7 (1983): 89–105; Claudio Guillén, "Notes toward the Study of the Renaissance Letter," in *Renaissance Genres: Essays on Theory, History, and Interpretation*, ed. Barbara Kiefer Lewalski (Cambridge, MA: Harvard University Press, 1986), 70–101; Erika Rummel, "Erasmus' Manual of Letter-writing: Tradition and Innovation," *Renaissance and Reformation* n.s., 13 (1989): 299–312; Gideon Burton, "From *Ars dictaminis* to *Ars conscribendi epistolis*: Renaissance Letter-Writing Manuals in the Context of Humanism," in *Letter-Writing Manuals and Instruction from Antiquity to the Present*, ed. Carol Poster and Linda C. Mitchell (Columbia: University of South Carolina Press, 2007), 88–101; Toon Van Houdt and Jan Papy, "Introduction," 1–13, Judith Rice Henderson, "Humanist Letter Writing: Private Conversation or Public Forum?" 17–38, and Charles Fantazzi, "Vives versus Erasmus on the Art of Letter Writing," 39–56, all in *Self-Presentation and Social Identification: The Rhetoric and Pragmatics of Letter Writing in Early Modern Times*, ed. Toon Van Houdt, Jan Papy, et al. (Leuven: Leuven University Press, 2002); Ian F. McNeely with Lisa Wolverton, *Reinventing Knowledge from Alexandria to the Internet* (New York: W. W. Norton, 2008), chapter 4, 119–59; Florike Egmond, "Correspondence and Natural History in the Sixteenth Century: Cultures of Exchange in the Circle of Carolus Clusius," in *Cultural Exchange in Early Modern Europe*, vol. 3, *Correspondence and Cultural Exchange in Europe 1400–1700*, ed. Francisco Bethencourt and Florike Egmond (Cambridge: Cambridge University Press, 2007), 104–42; Adam Mosley, *Bearing the Heavens: Tycho Brahe and the Astronomical Community of the Late Sixteenth Century* (Cambridge: Cambridge University Press, 2007), chapter 2; and idem, "Tycho Brahe's *Epistolae astronomicae*: A Reappraisal," in Van Houdt, Papy, et al., *Self-Presentation and Social Identification*, 449–68.

Regarding ongoing research on the Republic of Letters, mention should specially be made of the digital project at Stanford University, Mapping the Republic of Letters: Exploring Correspondence and Intellectual Community in the Early Modern Period (1500–1800), https://republicofletters.stanford.edu/ (the case studies so far completed for this project concern individuals who flourished in the seventeenth and eighteenth centuries); see also Anthony Grafton, "A Sketch Map of a Lost Continent: The Republic of Letters," *Republics of Letters: A Journal for the Study of Knowledge, Politics, and the Arts* 1, no. 1 (May 1, 2009), http://arcade.stanford.edu/journals/rofl/node/34 (which contains a rich bibliography of earlier writing on the subject of the Republic of Letters).

2. Ian Maclean, "The Medical Republic of Letters before the Thirty Years War," *Intellectual History Review* 18 (2008): 15–30, with an appendix at 29–30 listing published collections of medical letters from 1521 to 1626. The subject is further treated in idem, *Learning and the Market Place: Essays on the History of the Early Modern Book* (Leiden: Brill, 2009), especially in chapter 4, "The Diffusion of Learned Medicine in the Sixteenth Century through the Printed Book," 59–86,

with specific comments on medical letter collections at 60 and a slightly revised list of published collections at 84–86.

3. Candice Delisle, "Accessing Nature, Circulating Knowledge: Conrad Gesner's Correspondence Networks and His Medical and Naturalist Practices," *History of Universities* 23 (2008): 35–58, and eadem, "The Letter: Private Text or Public Place? The Mattioli-Gesner Controversy about the *aconitum primum*," *Gesnerus* 61 (2004): 161–76 (issue 3/4 of this volume of *Gesnerus* is devoted to early modern medical correspondence, but all the other articles discuss examples from the late seventeenth or eighteenth century).

4. Giovanni Manardo, *Epistolae medicinales in quibus multa recentiorum errata et antiquorum decreta reserantur* [Ferrara: Formis excussit Bernardinus de Odonino, 1521]; idem, *En postremum tibi damus, candide lector . . . epistolarum medicinalium libros XX. e quibus ultimi duo in hac editione primum accesserunt, una cum epistola iandudum desiderata, de morbis interioribus, quam utinam immatura morte non praeventus, totam absolvere potuisset. Eiusdem in Ioan. Mesue Simplicia & composita annotationes & censurae, omnibus practicae studiosis adeo necessariae, ut sine harum cognitione aegrotantibus recte consulere nemo possit* (Basel: apud Mich. Isingrinium, 1540); cited hereafter as Manardo, *Epistolarum medicinalium libri XX* [1540]).

5. Delisle, "Accessing Nature, Circulating Knowledge," 36; Ann M. Blair, *Too Much to Know: Managing Scholarly Information before the Modern Age* (New Haven: Yale University Press, 2010), 193, citing the studies on Zwinger of Carlos Gilly. For Clusius's correspondence and correspondents, see the Clusius Project of the Scaliger Institute, Leiden University, www.library.leiden.edu/special-collections/scaliger-institute/projects/clusius-project.html; Egmond, "Correspondence and Natural History in the Sixteenth Century"; and eadem, "Apothecaries as Experts and Brokers in the Sixteenth-Century Network of the Naturalist Carolus Clusius," *History of Universities* 23, no. 2 (2008): 59–91. See also Conrad Gesner, *Epistolarum medicinalium Conradi Gesneri, philosophi et medici Tigurini, libri III . . . Omnia nunc primum per Casparum Wolphium medicum Tigurinum, in lucem data* (Zurich: excudebat Christoph. Frosch., 1577).

6. On the continuous expansion in the sixteenth century of the accumulation and dissemination of descriptive information about the world of nature, see Brian Ogilvie, "The Many Books of Nature: Renaissance Naturalists and Information Overload," *Journal of the History of Ideas* 64 (2003): 29–40, and, more generally, idem, *The Science of Describing: Natural History in Renaissance Europe* (Chicago: University of Chicago Press, 2006). Harold J. Cook, *Matters of Exchange: Commerce, Medicine, and Science in the Dutch Golden Age* (New Haven: Yale University Press, 2007), argues for the significance of the exchange of descriptive information in and among the seventeenth-century Dutch Republic and its overseas territories for developments both in science and in commercial life.

7. See Delisle, "The Letter: Private Text or Public Place?," for a striking example of a dispute. On Mattioli and the "botanical republic," see also Paula Findlen, "The Formation of a Scientific Community: Natural History in Sixteenth-Century Italy," in *Natural Particulars: Nature and the Disciplines in Renaissance Europe*, ed. Anthony Grafton and Nancy Siraisi (Cambridge, MA: MIT Press, 1999), 369–400.

8. Gesner, *Epistolarum medicinalium . . . libri III*, bk. 1, [letter 1], 1r–2v: "Conradus Gesnerus Ioanni Cratoni a Crafftheim, S. Caes. Maiest. medico intimo." On Crato's anti-Paracelsianism, see Charles D. Gunnoe, Jr., and Jole Shackelford, "Johannes Crato von Krafftheim (1519–1585): Imperial Physician, Irenicist, and Anti-Paracelsian," in *Ideas and Cultural Margins in Early Modern Germany: Essays in Honor of H. C. Erik Midelfort*, ed. Marjorie Elizabeth Plummer, Robin B. Barnes, et al. (Farnham, Surrey: Ashgate, 2009), 201–16.

9. Desiderius Erasmus, *Opus de conscribendis epistolis*. The first edition authorized by Erasmus appeared in 1522. See Charles Fantazzi, "Introductory Note," in Desiderius Erasmus, *De conscribendis epistolis / On the Writing of Letters*, trans. and annotated Charles Fantazzi, in *Collected Works of Erasmus*, vol. 25, *Literary and Educational Writings: Volume 3*, ed. J. K. Sowards (Toronto: University of Toronto Press, 1985), 1–9, with the translation of Erasmus's treatise at 12–254. On the printing history, content, and influence of this work, see also Gerlo, "The *Opus de conscribendis epistolis* of Erasmus"; Rummel, "Erasmus' Manual of Letter-writing"; Henderson, "Erasmus on the Art of Letter-Writing"; and eadem, "Humanist Letter Writing."

10. Luigi Mondella, *Dialogi medicinales decem, nunc primum in lucem editi: in quibus multa & varia tum artis theoremata, tum historiae & experimenta doctissime explicantur* (Zurich: apud Froschouerum, 1551), Dialogus VIII, 93r: "iis enim proximis diebus ex Helvetiis mihi nunciatum est a Gesnero medico, viro quidem undecunque erudito, ac mihi amicissimo, quendam medicum senem nuperrime vita defunctum esse, in cuius librorum suppellectile inventae sunt singulares quaedam compositiones, longoque usu probatae, quas ille chymica arte invenerat, quarum ille, cum mihi libuerit, me participem facere pollicitus est.

"Lyc.: Cum primum illae ad te allatae erunt, ut mecum quoque eas communices precor, quo ita respublica Christiana in primis, universaeque deinde nationes hoc beneficio, et commodo iuvari, et gaudere possint." My thanks to Gianna Pomata for drawing my attention to this passage.

11. Gesner was the author of *Thesaurus Euonymi Philiatri de remediis secretis* (1552 and numerous subsequent early editions), which contains recipes for many chymically prepared medicines; see Charles Webster, "Conrad Gessner and the Infidelity of Paracelsus," in *New Perspectives on Renaissance Thought*, ed. John Henry and Sarah Hutton (London: Gerald Duckworth, 1990), 13–23, at 20–23.

12. Luigi Mondella, *Epistolae medicinales, variarum quaestionum, & locorum insuper Galeni difficilium expositionem continentes, omnibus qui veram artem exer-*

cere volunt apprime utiles. Eiusdem annotationes in Antonii Musae Brasavolae Simplicium medicamentorum examen (Basel: apud Mich. Isingrinium, 1543), consulted www.e-rara.ch/doi/10.3931/e-rara-3955, containing twenty-two letters; idem, *Epistolae medicinales, nunc ab ipso autore auctae et recognitae: in quibus variae & difficiles quaestiones utiliter tractantur: Galeni, atque aliorum medicorum loci obscuri & implicati illustrantur & explicantur: quae quidem omnia, omnibus verae & incorruptae medicinae studiosis tum utilissima, tum necessaria sunt. Eiusdem Annotationes* (Basel: apud Mich. Ising[rinium], [ca. 1550]), consulted www.e-rara.ch/doi/10.3931/e-rara-483, which includes the twenty-two letters printed in 1543 plus another fourteen (individual items bear dates from 1537 to 1549). Mondella's letters are also included in *Epistolae medicinales diversorum authorum nempe Ioannis Manardi . . . Nicolai Massae . . . Aloisii Mundellae . . . Io. Baptistae Theodosii . . . Ioan. Langii Lembergii* (Lyon: apud haeredes Iacobi Juntae, 1556; cited hereafter as *Epistolae medicinales diversorum authorum* [1556]), pp. 320–402. The Lyon collection includes the same thirty-six letters as the last Basel edition. Mondella seems to have died in the 1560s, as the preface to his *Theatrum Galeni* (Basel: per Eusebium Episcopium et Nicolai fratris heredes, 1568), by Giovanni Battista Mondella, α2r, says that Luigi Mondella's recent death occurred around the same time as that of the humanist and Aristotelian scholar Vincenzo Maggi (which was probably in 1564).

13. Zurich, Zentralbibliothek ms C 50a (227) Brief an Konrad Gessner, Luigi Mondella, Professor der Medizin und Botanik, gest. 1553 Brixiae [Brescia], 3. Non. Oct. 1552 [?] [5.10.1552], and ms C 50a (228) Brief an Konrad Gessner, Luigi Mondella, Professor der Medizin und Botanik, gest. 1553 Brixiae [Brescia], 3. Cal. Oct.1552 [?] [29.9.1552] (my thanks to Dr. Alexa Renggli of the Department of Manuscripts, Zentralbibliothek Zurich for arranging for me to receive photocopies of these letters); Conrad Gesner, *Historiae animalium liber III, qui est de avium natura* (Zurich: apud Christoph. Froschouerum, 1555). At pp. 219–20: "De attagene . . . Est autem eadem Germanorum Haselbun, id est corylorum gallina . . . quod et in Italia ipse audivi, et ex icone francolini Venetiis dicti, quam doctissimus medicus Aloisius Mundella ad me misit."

At p. 338: "De coturnice. Coturnicem eam esse avem, cuius hic figura cernitur, ut homines docti nostra aetatis omnes consentiunt, et copiose in epistolis suis Aloisius Mundella medicus approbavit: ita literatores quidam repugnant, eam quae hodieque vulgo coturnix apud Italos dicatur veterum quoque coturnicem esse putantes, sed horum error crassior est, quam ut nos in eo refellendo tempus collocare conveniat, et iam a doctissimo viro Mundella satis est reprehensus."

At pp. 655–56: "De perdice maiore, quam Itali vulgo coturnicem vocant . . . Coturnicem quae vulgo appellatur, non esset forte admodum absurdum gallinaginis aliquam speciem esse putare, sive avis Numidicae, quam quidam deserti gallinam vocant, Aloisius Mundella" (several other references to Mondella's account follow on the same page). See Mondella, *Epistolae medicinales* (ca. 1550), epistola 6 (dated at end 1540), pp. 106–23.

14. On heresy and the Counter-Reformation in sixteenth-century Brescia, see Massimo Firpo, "The Italian Reformation," in *A Companion to the Reformation World*, ed. R. Po-chia Hsia (Oxford: Blackwell, 2004), 169–84, at 172; idem, *Riforma protestante ed eresie nell'Italia del Cinquecento: Un profilo storico* (Rome: Laterza, 1993), chapter 2, "La 'porta' della Riforma: Venezia," 11–28, especially 22–23; Christopher F. Black, *Church, Religion and Society in Early Modern Italy* (Basingstoke, UK: Palgrave Macmillan, 2004), 11–18; Christopher Cairns, *Domenico Bollani, Bishop of Brescia: Devotion to Church and State in the Republic of Venice in the Sixteenth Century* (Nieuwkoop: B. DeGraaf, 1976), especially chapter 8; *Storia di Brescia*, 5 vols. ([Brescia]: Morcelliana, 1963–64), 2:437–73. Mondella, *Epistolae medicinales* (ca. 1550), epistola 26 (dated at end 1543), pp. 364–75, responds to the accusation that he prescribed the same medicine for patients of every social condition (i.e., regardless of the supposedly greater physical delicacy of the upper classes). At pp. 370–71: "Utinam charitas Christiana nobiscum habitaret, utinam nos mutuo amore prosequeremur . . . Sed quoniam ea longe a cordibus nostris recessit, omniaque in peius ruere non cessant, et non nisi quaedam vana illius religio et superstitio a nobis excolitur et veneratur, itaque lege cautum est humana, ut qui se nobilem appellari velit, illi omnino necesse sit, ut neque ipse, aut parentes eius, aut etiam atavi, ulli unquam sedentariae vacaverint arti, aut etiam vacent, sed ociosam vitam degat, ac delicate vivat: alioqui nec clarum illum ullo modo esse posse, ne illis fungi honoribus et dignitatibus, quibus caeteri omnes, quos isti appellant nobiles: denique et privata utatur tum victus, tum vestitus ratione, pro sua, suaeque familiae sustentatione: civitate, meo quidem iudicio, Christiana indignum. Quoniam igitur Christus rex regum, et generis claritate caeteros omnes praecellens, fabrilem artem pertractare non veritus est, fabrique filius nuncupari, civilibus ergo non potietur honoribus? Sacris non ascribetur collegis? Urbis dignitatibus ullis non gaudebit? Quare respicendum iam est, et pulvis ex oculis nostris excutiendus, neque amplius dicendum, pauperes et ignobiles alio et alio curandos esse quam divites et nobiles modo, ut medicorum nonnulli, et alii diversi generis viri existimant et praedicant, quod charitas atque amor longe ab illorum absit cogitationibus."

15. Mondella, *Epistolae medicinales* (ca. 1550), epistola 20 (to Rev. Angelo Castiglione, on his deafness, dated at end 1542), pp. 284–300. At p. 300: "Vale, et me, quod facis, in Christo ama, saepiusque de te ac tuis rebus scribas, rogo: nihil enim mihi gratius continere potest, quam de prospera valetudine tua intelligere, cum amicorum habeam neminem, quem plus ac te amem et diligam." On Castiglione, see Albano Biondi, "Castiglione (Castiglioni, Castilionacus), Angelo," *DBI*, www.treccani.it/enciclopedia/angelo-castiglione_(Dizionario-Biografico)/. All of Gesner's works were placed on the 1559 *Index librorum prohibitorum*, but this action was taken after Mondella's known correspondence with him.

16. Lorenz Scholz von Rosenau, *Consiliorum medicinalium, conscriptorum a praestantissimis atque exercitatissimis nostrorum temporum medicis liber singularis*

(Frankfurt: apud Andreae Wecheli haeredes, Claudium Marnium et Ioannem Aubrium, 1598]), preface (dated at end March 26, 1598), [5v]; idem, *Epistolarum philosophicarum, medicinalium, ac chymicarum a summis nostrae aetatis philosophis ac medicis exaratarum volumen; Opus, cum ob remediorum saluberrimorum copiam, cum ob variam doctrinam, ac varii generis difficilium et obscurarum quaestionum explicationem, non solum medicis verumetiam philosophis admodum utile et necessarium* (Frankfurt: apud Andreae Wecheli haeredes, Claudium Marnium et Ioannem Aubrium, 1598; cited hereafter as Scholz, *Epistolae* [1598]), preface (dated at end March 30, 1598), *4r. On Scholz (1552–1599), who was a collector of plants as well as manuscripts and the founder of a botanic garden at Breslau, see *ADB* 32:229–30, www.deutsche-biographie.de/sfz3154.html.

17. Scholz, *Consiliorum medicinalium . . . liber singularis*, 3r–v: "Duo autem praecipue horum consiliorum fines sunt; unus cum iis, qui longius absunt, si nostrum de sua praesente in valetudine iudicium requirunt, gratificamur; alter, cum in chronicis et diuturnis, ac periodicis morbis, de earum rerum, quae impedire, et intercipere novas invasiones possunt, observatione aliquid monemus."

18. Scholz, *Epistolae* (1598), *3r–v: "libet mihi nonnihil praefationis loco de usu epistolarum in medium adferre. Harum autem non minorem, quam consiliorum medicinalium utilitatem esse, nulla, ut ego quidem iudico, probatione egere videtur. Vetus est ratio de problemate proposito interrogare, disputare, ac percunctanti et sciscitanti respondere, quodque responsum est ad examen revocare. Tales epistolas, suo quodam iudicio, rhetores in scholis dialecticas, erotematicas, et didascalicas appellare consueverunt. Estque hoc genus epistolarum familiare quidem, at non vulgo, sed eruditorum saltem generi proprium. Hi enim ultro citroque missis literis de variis rebus, si quando, vel omnino dubitant, vel in sententiis inter se non prorsus conveniunt, disputare solent. Huius vel in Cicerone exemplum illud esse potest, cuius initium 'Illuseras heri inter schyphos, &c' . . . Tales sunt, ut exempla alia non accersam, duae illustres ille de eloquentia scriptae epistolae contrariae, una Francisci [*sic*] Pici Mirandulani Comitis, alter Melanchthonis nostri, in scholis passim notae et celebres. Critici recentiores, quorum pene nimis est ferax nostrum seculum, quot non epistolas de certorum locorum in autoribus lectione varia, ac restitutione ediderunt? Id si in minima, ac saepe nugatoria re, imo vocula, fieri concessum est, quid nobis in re medica, si quid occurrat, quod nobis dubitatione iniiciat, faciendum esse existabimus? Averrhoes de se ipso testatur, quod saepenumero per epistolas amicorum absentium, quibuscunque de rebus visum fuisset, sententias explorare solitus fuerit. Extant quinetiam epistolae multae cum veterum theologorum Graecorum et Latinorum, tum aliorum quoque in primis vero duodecim illae nobiles Platonis, scriptoris gravissimi, ut taceam de iis quas princeps noster Hippocrates scripsit, quoque in operibus eius leguntur." The list of recent authors of *epistolae medicinales* follows at *3v–*4r: Giovanni Manardo, Niccolò Massa, Luigi Mondella, Giovanni Battista Teodosi, Johann Lange,

Conrad Gesner, Pietro Andrea Mattioli, Giulio Alessandrini, Orazio Augenio, Andreas Dudith, Girolamo Mercuriale, Thomas Erastus, and Johann Crato von Krafftheim. On the debate between Giovanni Pico della Mirandola and Ermolao Barbaro on the respective merits of philosophy and rhetoric, see Jill Kraye, "Pico on the Relationship of Rhetoric and Philosophy," in *Pico della Mirandola: New Essays*, ed. M. V. Dougherty (Cambridge: Cambridge University Press, 2008), 13–36. For the contribution attributed to Melanchthon but actually written by Franz Burchard (Franciscus Vinariensis), see Erika Rummel, "*Epistola Hermolai nova ac subditicia*: A Declamation Falsely Ascribed to Philip Melanchthon," *Archiv für Reformationsgeschichte* 83 (1992), 302–5; she argues that Burchard's *Epistola* was edited by Melanchthon and reprinted with some editions of Melanchthon's treatise on rhetoric before being included as the latter's own work in the edition of his letters published in Wittenberg in 1565.

19. For the letters attributed to Hippocrates, see Hippocrates, *Pseudepigraphic Writings*, ed. and trans. Wesley D. Smith (Leiden: Brill, 1990), and for the sixteenth-century printed editions, ibid., 35. For early fifteenth-century humanist translations into Latin of letters ascribed to Hippocrates and his correspondents, see Pearl Kibre, *Hippocrates Latinus* (New York: Fordham University Press, 1985), 158–62. The letters were also translated by Fabio Calvi and included in his printed Latin *Opera* of Hippocrates (Rome, 1525). On ancient medical letters more generally, see D. R. Langslow, "The *Epistula* in Ancient Scientific and Technical Literature, with Special Reference to Medicine," in *Ancient Letters: Classical and Late Antique Epistolography*, ed. Ruth Morello and A. D. Morrison (Oxford: Oxford University Press, 2007), 211–34.

20. See Jole Agrimi and Chiara Crisciani, *Les Consilia médicaux*, trans. Caroline Viola, Typologie des Sources du Moyen Age Occidental 69 (Turnhout, Belgium: Brepols, 1994), and Gianna Pomata, "*Praxis Historialis*: The Uses of *Historia* in Early Modern Medicine," in *Historia: Empiricism and Erudition in Early Modern Europe*, ed. Gianna Pomata and Nancy G. Siraisi (Cambridge, MA: MIT Press, 2005), 105–46, especially 122–37.

21. See, for example, Marilyn Nicoud, "Expérience de la maladie et échange épistolaire: les derniers moments de Bianca Maria Visconti (mai–octobre 1468)," *Mélanges de l'Ecole française de Rome (Moyen Age)* 112, no. 1 (2000): 311–458. For a mid-twelfth-century example of an exchange of correspondence between a patient and his doctor, see Giles Constable, ed., *The Letters of Peter the Venerable*, 2 vols. (Cambridge, MA: Harvard University Press, 1967), 1:379–83, letters 158a and 158b, discussed in Nancy G. Siraisi, *Medieval and Early Renaissance Medicine* (Chicago: University of Chicago Press, 1990), 115–18. Michael R. McVaugh, *Medicine before the Plague: Practitioners and Their Patients in the Crown of Aragon, 1285–1345* (Cambridge: Cambridge University Press, 1993), chapter 1, describes medicine at the Aragonese court, drawing, inter alia, upon royal correspondence that includes a number of letters from successive rulers to medical practitioners.

22. On concepts relating to witnessing and eyewitnessing in Renaissance medicine, see Ian Maclean, "Evidence, Logic, the Rule and the Exception in Renaissance Law and Medicine," *Early Science and Medicine* 5 (2000): 227–57; on *observationes*, see Gianna Pomata, "*Observatio* ovvero *historia*: Note su empirismo e storia in età moderna," *Quaderni Storici* 31 (1996): 173–98.

23. On the early humanist dialogue as a vehicle for the presentation of different points of view (and the role of correspondence in giving advice about the composition of one such dialogue), see Helene Harth, "Niccolò Niccoli als literarischer Zensor: Untersuchungen zur Textgeschichte von Poggios 'De Avaritia,'" *Rinascimento*, ser. 2, 7 (1967): 29–53, especially 52–53. I owe this reference to Anthony Grafton.

24. Manardo, *Epistolarum medicinalium libri XX* (1540), I.I, p. 3: "Modus autem hic scribendi per epistolas, praeter id quod vitari vix potest, interpellantibus amicis, non est novus. Archigenes, Galeno teste, undecim libros epistolarum medicinalium scripsit: et Themison, Paulo teste, decem. Consilia etiam a recentioribus vocata, non aliud certe sunt, quam epistolae."

25. In the collection of the library of the New York Academy of Medicine, Scholz, *Consiliorum medicinalium . . . liber singularis* (1598) is bound with idem, *Epistolae* (1598); of the two works the former has evidently been much more heavily used.

26. See Clough, "The Cult of Antiquity"; Fumaroli, "Genèse de l'épistolographie classique"; Henderson, "Erasmus on the Art of Letter-Writing"; Burton, "From *Ars dictaminis* to *Ars conscribendi epistolis*"; and Fantazzi, "Vives versus Erasmus on the Art of Letter Writing."

27. Desiderius Erasmus, *Opus de conscribendis epistolis, ex postrema autoris recognitione emendatius editum* (Cologne: Ioannes Gymnicas excudebat, 1537), p. 414: "Disputatoriae genus. Est epistolarum genus non infrequens inter eruditos, quo studiorum suorum inter se agunt commercia, quum aut sciscitantur de re quapiam, aut respondent sciscitanti: aut si qua de re parum convenit disputant." The English translation of the treatise in Erasmus, *Collected Works of Erasmus*, vol. 25, *Literary and Educational Writings: Volume 3; De conscribendis epistolis*, 254, renders Erasmus's "disputatoriae genus" as "letters of discussion." Juan Luis Vives, *De conscribendis epistolis*, ed. and trans. Charles Fantazzi (Leiden: Brill, 1989), 24–26: "Additae sunt postea consolatoriae, conciliatoriae, praeceptoriae, disputatoriae de omni argumento philosophiae, iuris, antiquitatis, omnium denique disciplinarum, atque earum rerum quae de scripto inter maxime praesentes agerentur. Sic a Platone de philosophia scribitur ad Dionysium et alios; a Seneca ad Lucilium; a Hieronymo, Ambrosio, Augustino, Cypriano de rebus sacris ad varios. Citantur Catonis Censorii et multorum iurisconsultorum libri de quaesitis aut responsis per epistolam." Justus Lipsius, *Principles of Letter Writing: A Bilingual Text of Justi Lipsi Epistolica Institutio*, ed. and trans. R. V. Young and M. Thomas Hester (Carbondale: Southern Illinois University Press, 1996), 20: "DOCTAM dico, quae ea

quae ad scientiam aut sapientiam, continet: et res non epistolae, epistolae veste velat." I am grateful to Anthony Grafton for the references to Vives and Lipsius.

28. See Vivian Nutton, "The Pleasures of Erudition: Mercuriale's *Variae lectiones*," in *Girolamo Mercuriale: Medicina e cultura nell'Europa del Cinquecento*, ed. Alessandro Arcangeli and Vivian Nutton (Florence: Leo S. Olschki Editore, 2008), 193–96, which points out that "variae lectiones" and similar expressions were used not only, as in the modern philological sense, for variant readings in different manuscripts or printed texts of the same work, but also for collections of emendations to or interpretations of passages in ancient authors.

29. *Epistolae medicinales diversorum authorum* (1556); however, the letters of each of the authors in that collection had also appeared separately before 1556 (see Maclean, "Medical Republic of Letters," 29).

30. Pietro Andrea Mattioli (1501–1578), *Epistolarum medicinalium libri quinque* (Lyon: apud Caesarem Farinam, 1564), the first edition of which appeared in 1561 (Maclean, "Medical Republic of Letters," 29). Orazio Augenio, *Epistolarum medicinalium tomi tertii libri duodecim* . . . [Frankfurt: apud heredes Andreae Wecheli, Claudium Marnium, & Ioannem Aubrium, 1600], *3v: "Hae caussa extiterunt, quam ob rem ego nunc minime promam, quas supra recensui lucubrationes meas sed pro illis tertium hunc Epistolarum medicinalium tomum emittam," consulted http://opacplus.bsb-muenchen.de/search?oclcno=159907655.

31. Vettor Trincavella, *Consiliorum medicinalium libri III. Epistolarum medicinalium libri III* (Venice: apud Camillum Borgominerium, 1586), with a dedicatory preface by Trincavello's son Bernardo. At α2r–v: "Itaque hinc factum est, cum intelligerem alia quoque illius opera, quae apud me extant adhuc, expectari quam avidissime, ut ego hoc tempore statuerim Epistolas et Consilia medica in lucem proferre." Wolf gave some account of his collecting and editing procedures in his dedicatory epistle in Gesner, *Epistolarum medicinalium . . . libri III*, α2r–β.

32. Johann Crato, *Consiliorum et epistolarum medicinalium, liber, Ex collectaneis clarissimi viri Dn. Petri Monavii Vratislavensis quondam medici Caesarei selectus, et nunc primum a Laurentio Scholzio . . . in lucem editus* (Frankfurt: apud Andreae Wecheli haeredes, Claudium Marnium & Joannem Aubrium, 1591; cited hereafter as Crato, bk. [1] [1591]). Books after the first added "et aliorum praestantissimorum medicorum." Books 2 and 3 followed from the same publisher in 1592, and books 4 and 5 in 1593; books 6 and 7 appeared from the house of Wechel in 1611; books 2–5 are cited hereafter as Crato et al., bk. 2 (1592), and so forth. On Crato, see Gunnoe and Shackelford, "Johannes Crato von Krafftheim."

33. On the circulation of some letters in manuscript beyond the original recipient, see Henderson, "Humanist Letter Writing," and Delisle, "The Letter: Private Text or Public Place?" For the preference of some sixteenth- and seventeenth-century authors for "scribal publication"—especially for works that were designed for a small, select audience, appealed to a particular patron, or involved political or

social risk—see Harold Love, *The Culture and Commerce of Texts: Scribal Publication in Seventeenth-Century England* (Amherst: University of Massachusetts Press, 1998; original edition Oxford: Oxford University Press, 1993), and Brian Richardson, *Manuscript Culture in Renaissance Italy* (Cambridge: Cambridge University Press, 2009); both works are primarily concerned with vernacular writings and mostly with literary texts. For other categories of books that continued to circulate in manuscript in early modern Italy, see also Federico Barbierato, "Tra attori e inquisitori: Manoscritti, commedia dell'arte e diffusione delle conoscenze magiche nella Venezia del Seicento," in *I luoghi dell'immaginario barocco: Atti del convegno di Siena, 21–23 ottobre 1999*, ed. Lucia Strappini, 341–53, and Yaacob Dweck, *The Scandal of Kabbalah: Leon Modena, Jewish Mysticism, Early Modern Venice* (Princeton: Princeton University Press, 2011), chapter 1, "Hebrew Manuscripts in the Age of Print," 47–76. I owe the last two references to Andrew Berns.

34. Philippe Tingus, letter of dedication, in *Epistolae medicinales diversorum authorum* (1556): "Ut rei medicae studiosis quam plurimam possent, adiumenti afferent vix ulli tamen hactenus operam suam melius utilius videntur collocasse, quam qui conscriptis eximiis aliquot Epistolarum (ut vocant) Medicinalium libris, densissimas quasque artis tenebras mira quadam ingenii dexteritate discusserunt. De hoc itaque Epistolarum genere cum non semel (uti sit in familiari amicorum colloquio) mentio nobis incidisset, memini te earum utilitatem magnopere commendare, et pro singulari tua facundia miris in coelum laudibus efferre: eam tamen ipsam longe uberiorum futuram affirmare, si quae hactenus sparse legebantur, et veluti dissecta naufragii tabulae, ex totidem codicibus, quot earum sunt autores, non sine gravi Lectoris molestia petendae erant, in unum quasi corpus contraherentur, ita ut unico volumine comprehensae, universae simul studiosorum oculis subijcerentur. Quod tuum consilium cum non possemus non magnopere approbare, apud patronos nostros, Iacobi Iuntae haeredes, multis nobis nominibus observandos, instare non destitimus, ne tam commodem de rebus mortales in bene merendi occasionem e manibus amitterent." In fact, the volume brought together several previously published collections.

Nicolaus Reusner, "Praefatio," in Johann Lange, *Epistolarum medicinalium volumen tripartitum, denuo recognitum, et dimidia sui parte auctum* (Frankfurt: apud heredes Andreae Wecheli, Claudium Marnium & Joannem Aubrium, 1589), βr: "praeclarum istud Epistolarum Medicinalium opus, studio et labore tuo incredibili denuo recognitum, et magna accessione locupletatum, iterum lucem videre voluisti: et quidem lucem accipere novam in praecipua Germaniae nostrae officina Wecheliana: quae et typorum elegantia et operarum industria, et operum denique ipsorum delectu singulari, laudem et commendationem hactenus meretur eximiam."

Theophilus Mader, letter of dedication, in Thomas Erastus, *Disputationum et epistolarum medicinalium volumen doctissimum. Nunc recens in lucem editum, opera* *et studio Theophili Maderi philosophiae ac medicinae doctoris ac physices in Academia Heydelbergensi Professoris ordinarii* (Zurich: apud Ioannem Wolphium, typis Frosch., 1595), *3r: "Caeterum quo maiorem etiam apud medicinae studiosos inirem gratiam, epistolas Erasti medicas, et eas, quibus theses de saporibus, morbis totius substantiae, de putredine, deque contagio explicantur magis et defenduntur, et alias diversi argumenti adiicere volui. Continent enim gravissimarum et obscurissimarum in arte nostra quaestionum declarationem eruditam, solidam, et perspicuam. In quarum collectione me non parum esse adiutum a matrona nobilissima ac lectissima uxore olim D. Erasti." Consulted http://opacplus.bsb-muenchen.de/search?oclcno=638260752.

35. Maclean, "Medical Republic of Letters," 21–22.

36. Mattioli, *Epistolarum medicinalium libri quinque.* In another disciplinary field, Tycho Brahe's *Epistolae astronomicae* included both sides of his correspondence with Landgrave Wilhelm of Hesse Kassel and his associates (see Mosley, *Bearing the Heavens*, chapter 2).

37. For an insightful study of the regional distribution of and professional and personal ties within Gesner's correspondence network, see Delisle, "Accessing Nature, Circulating Knowledge."

Chapter 1 · Contexts and Communication

1. See Bruno Caizzi, *Dalla posta dei re alla posta di tutti: Territorio e communicazioni in Italia dal XVI secolo all'Unità* (Milan: FrancoAngeli, 1993), 211–49; Luciano De Zanche, *Tra Costantinopoli e Venezia: Dispacci di stato e lettere di mercanti dal Basso Medioevo alla caduta della Serenissima*, Quaderni di Storia Postale 25 (Prato: Istituto di studi storici postali, 2000), 5–14; and Filippo de Vivo, *Information and Communication in Venice: Rethinking Early Modern Politics* (Oxford: Oxford University Press, 2007), especially 52–53.

2. On the *peregrinatio medica*, see Ole Peter Grell, Andrew Cunningham, and Jon Arrizabalaga, eds., *Centres of Medical Excellence? Medical Travel and Education in Europe, 1500–1789* (Farnham, Surrey: Ashgate, 2010).

3. Nicolò Massa, *Epistolae medicinales, et philosophicae* [Venice: apud Franciscum Bindonum, & Mapheum Pasinum, 1550] (epistolae 5 and 6, to the Paduan professor Antonio Fracanzano, 51v–60r, discuss anatomy); idem, *Epistolarum medicinalium tomus primus [tomus alter]* (Venice: ex officina Stellae Jordani Zilleti, 1558). The first book was reissued as part of the collection *Epistolae medicinales diversorum authorum* (1556). On Massa (1489–1569), see L. Roscioni, "Massa, Niccolò," *DBI*, www.treccani.it/enciclopedia/niccolo-massa_(Dizionario-Biografico)/; L. R. Lind, *Studies in Pre-Vesalian Anatomy* (Philadelphia: American Philosophical Society, 1975), 167–253; Richard Palmer, "Nicolò Massa, His Family and His Fortune," *Medical History* 25 (1981): 385–410. On medical teaching at Venice, see Richard Palmer, *The Studio of Venice and Its Graduates in the Sixteenth Century* (Trieste:

Edizioni Lint, 1983), especially 44–47. The *studium* of Venice did not offer a complete formal medical curriculum; many of those who took medical degrees at Venice did so after medical studies elsewhere. Important anatomical demonstrations and lectures were also given annually in Venice.

4. Massa, *Epistolarum medicinalium [tomus alter]* (1558), table of contents:

"IIII. Expositio definitionis febris positae ab Avicenna primo Fen primae quarti Canonis, ad Laurentium Massam;

V. Expositio secundae partis primi capitis Fen primae quarti canonis Avicennae ad eundem.

VI. Expositio divisionis partis tertiae primi capitis primae Fen quarti canonis Avicennae ad eundem.

VII. De loco ubi putrescunt humores in febribus intermittentibus ad eundem."

On Lorenzo Massa's later reputation as being medically well informed, see Palmer, "Nicolò Massa," 407.

5. Palmer, "Nicolò Massa," 398–99. On Heinrich Stromer of Auerbach—rector of Leipzig University, 1508; graduated in medicine there, 1511; dean of the medical faculty, 1523—see Peter G. Bietenholz et al., *Contemporaries of Erasmus* (Toronto: University of Toronto Press, 1985), 291–92, s.v. "Stromer," and *ADB* 1 (1875): 638, www.deutsche-biographie.de/sfz45425.html#adb, s.v. "Auerbach."

6. For a bibliography of Massa's writings and an account of his work on syphilis, see Palmer, "Nicolò Massa," 391–94.

7. Massa, *Epistolae medicinales* (1550), 161v–167[misprint for 165]r: "Ad excellentissimum Henricum Stromerum Auribachum Epistola XXXIII de mundi creatione" (dated from Venice, 1543); ibid., 167[misprint for 165]r–168r: "Ad excellentissimum Henricum Stromerum Auribachium Epistola XXXIIII, de animi immortalitate" (dated from Venice, 1542). In *Epistolae medicinales diversorum authorum* (1556), these letters appear at pp. 316–18 and pp. 318–20.

8. Palmer, "Nicolò Massa," 398–99, noting that the elder Massa's religious identity seems to have remained firmly Catholic, and that Apollonio Massa's medical career in Venice was successful.

9. For a summary of the views of Pomponazzi (who taught at Padua, 1488–96 and 1499–1509) and Nifo on the soul and related controversies in the period ca. 1490–ca. 1525, see Eckhard Kessler, "The Intellective Soul," in *The Cambridge History of Renaissance Philosophy*, ed. Charles B. Schmitt et al. (Cambridge: Cambridge University Press, 1988), at 495–507.

10. For analytic historiographical overviews of late twentieth-century work on sixteenth-century Italian religious history, see Ann Jacobson Schutte, "Periodization of Sixteenth-Century Italian Religious History: The Post-Cantimori Paradigm Shift," *Journal of Modern History* 61 (1989): 269–84, and John Martin, "Recent Italian Scholarship on the Renaissance: Aspects of Christianity in Late Medieval and Early Modern Italy," *Renaissance Quarterly* 48 (1995): 593–610, especially 598–606.

11. On Duno and the Locarnese exiles, see Albert Chenou, "Taddeo Duno et la Réforme à Locarno," *Archivio Storico Ticinese* 71 (1971): 237–94; for Duno's medical letters, see *Thaddaei Duni . . . et Francisci Cigalini, Joannisque Pauli Turriani . . . item Hieronymi Cardani . . . disputationum per epistolas liber unus per quam utilis* (Zurich: per Andream et Jacobum Gesneros fratres [1555]).

12. For the midcentury Venetian measures against heresy, see Richard Palmer, "Physicians and the Inquisition in Sixteenth-Century Venice: The Case of Girolamo Donzellini," in *Medicine and the Reformation*, ed. Ole Peter Grell and Andrew Cunningham (London: Routledge, 1993), 118–33, especially 118–20. On religious conditions in Venice more generally, see John Jeffries Martin, *Venice's Hidden Enemies: Italian Heretics in a Renaissance City* (Los Angeles: University of California Press, 1993), together with the review of this work by Massimo Firpo, *Journal of Modern History* 68 (1996): 218–21.

13. Crato, *Consiliorum et epostolarum [libri]*. Edited by Scholz, books 1–7 (1591–1611); Scholz, *Epistolae* (1598).

14. Thomas Erastus, *Disputationum et epistolarum medicinalium volumen doctissimum. Nunc recens in lucem editum, opera et studio Theophili Maderi philosophiae ac medicinae doctoris ac physices in Academia Heydelbergensi Professoris ordinarii* (Zurich: apud Ioannem Wolphium, typis Frosch., 1595). On Erastus (1524–1583), see Ruth Wesel-Roth, "Erast (Lüber), Thomas," *NDB* 4 (1959): 560, www.deutsche-biographie.de/sfz23271.html.

15. Vivian Nutton, "The Reception of Fracastoro's Theory of Contagion: The Seed That Fell among Thorns," *Osiris* 6 (1990): 196–234, at 221–25.

16. On Paterno, see Jacopo Filippo Tomasini, *Gymnasium Patavinum . . . libris V comprehensum* (Udine, 1654), pp. 292–93; on Capodivacca's career and writings, see Giuliano Gliozzi, "Capodivacca (Capivaccio, Capivacceus), Girolamo," *DBI*, www.treccani.it/enciclopedia/girolamo-capodivacca_(Dizionario-Biografico)/. On the status of the respective professorial chairs and the rivalry between Capodivacca and Paterno concerning them, see Nancy G. Siraisi, *Avicenna in Renaissance Italy* (Princeton: Princeton University Press, 1987), 110–11.

17. Erastus, *Disputationum et epistolarum medicinalium volumen*, Disputatio 13, first series of foliation, 18r–24r: "respondente M. Timotheo Madero Helvetio. De contagio." For the graduation of Theophilus and Timotheus Mader as masters of arts, see Gustav Toepke, ed., *Die Matrikel der Universität Heidelberg von 1386 bis 1662: Zweiter Theil von 1554 bis 1662* (Heidelberg, 1886), http://digi.ub.uni-heidelberg.de/diglit/matrikel1554/0396/ocr, 464. Theophilus Mader's role as the editor of Erastus's *Disputationum et epistolarum medicinalium volumen* is indicated in the full title of that work. Thomas Erastus, *De causa morborum continente tractatus* (Basel: Pietro Perna, 1572), published with idem, *Disputationum de nova Philippi Paracelsi medicina pars tertia* ([Basel:] Pietro Perna, 1572).

18. Erastus, *Disputationum et epistolarum medicinalium volumen*, epistola 17 (September 22, 1574), second series of foliation, 50r–54v: "Ad D. Hieronymum

Capivaccium. Morbos quosdam sua essentia, contagiosos esse. 2. Tria tantum morborum genera esse, nec his posse addi morbos totius substantiae. 3. Utrum spiritus putrescere possint." At 50v: "Accepi a D. Theodorico Simelbeckero nostro, tuas adversus theses nostras de contagio obiectiones, vir clarissime omnique officiorum genere colendissime, una cum cogitationibus longe eruditissimis D. Paterni, die 20 Sept. quae mihi quovis auro gratiores, vel hoc nomine fuerunt, quod aditum mihi ad tantorum virorum familiaritatem fecerunt. Scribit enim Theodoricus noster de tua excellentia, in hunc modum. Excellentissimus D. Doct. Hieronymus de Capitavaccae, nobilis Patavinus, practices primarius professor, theses magnopere commendavit, mihique suum de eis iudicium dictavit. Cupit tecum de eisdem per literas, cum aliter non possit, conferre: ac petit, ut proxima occasione, rescribas." Theodoric Simmelbecker had matriculated at Heidelberg in 1561 (Toepke, *Die Matrikel der Universität Heidelberg . . . von 1554 bis 1662*, p. 27, no. 21); his name does not appear in the index of Antonio Favaro, ed., *Atti della nazione germanica artista nello studio di Padova*, 2 vols. (Venice, 1911–12), or of Lucia Rossetti and Giorgetta Bonfiglio-Dosia, eds., *Matricula nationis Germanicae artistarum in gymnasio Patavino, 1553–1721* (Padua: Editrice Antenore, 1986). For a full list of Erastus's letters and correspondents and a bibliography of his medical writings, see Charles D. Gunnoe, "Thomas Erastus in Heidelberg: A Renaissance Physician during the Second Reformation, 1558–1580" (PhD diss., University of Virginia, 1998), 381–401, 408–11.

19. Erastus, *Disputationum et epistolarum medicinalium volumen*, epistola 18 (September 24, 1574), second series of foliation, 54v–59v: "Ad D. Bernhardum Paternum. Continet defensionem suarem thesen de contagio et contagiosis morbis. 2. Spiritus utrum putrescere nati sint"; epistola 19 (February 4, 1575), second series of foliation, 59v–65r: "Ad D. Bernhardum Paternum. Eiusdem, cum praecedenti, argumenti epistola est, et praeterea an omne pestilens morbi genus, putridum sit"; epistola 20 (February 8, 1575), second series of foliation, 65r–70r: "Ad D. Hieron. Capivaccium. Utrum rectius medici de morbis ita disserant, ut logici solent, in abstracto eos considerantes, an ut physici solent, qui non sine subiecto et causis, eos contemplantur. Respiciunt vero fere omnia de ea quae in superioribus sunt tractata, uti etiam qua de spirituum putredine in fine dicuntur."

20. Scholz, *Epistolae* (1598), epistola 269 (Erastus to Capodivacca, September 8 1575), cols. 482–503 (also in Crato et al., bk. 4 [1593], pp. 268–315). At cols. 500–501: "Quare figmenta sunt, quae horum morborum architectus Fernelius, et Argenterius faber commenti sunt."

21. Jean Fernel, *De abditis rerum causis libri duo* (Paris, 1548) and numerous later editions; idem, *De naturali parte medicinae libri septem* (Paris, 1542), subsequently incorporated as the physiological section of his *Medicina* (Paris, 1554) and *Universa medicina* (Paris, 1567). Numerous editions of *Universa medicina* continued to be published until the late seventeenth century. See also James J. Bono, *The Word of God and the Languages of Man* (Madison: University of Wisconsin Press,

1995); Linda Deer Richardson, "The Generation of Disease: Occult Causes and Diseases of the Total Substance," in *The Medical Renaissance of the Sixteenth Century*, ed. A. Wear et al. (Cambridge: Cambridge University Press, 1985), 175–94; and Hiro Hirai, "Ficin, Fernel et Fracastor autour du concept de semence: Aspects platoniciens de *seminaria*," in *Girolamo Fracastoro fra medicina, filosofia e scienze della natura*, ed. Alessandro Pastore and Enrico Peruzzi (Florence: Olschki, 2006), 245–60.

22. On Argenterio, see Felice Mondella, "Argenterio, Giovanni," *DBI*, www .treccani.it/enciclopedia/giovanni-argenterio_(Dizionario-Biografico)/; Nancy G. Siraisi, "Sixteenth-Century Medical Innovation between Princely Patronage and Academic Controversy: The Case of Giovanni Argenterio," *Osiris* 6 (1990): 162–80; and eadem, "Disease and Symptom as Problematic Concepts in Renaissance Medicine," in *Res et verba in der Renaissance*, ed. Ian Maclean and Eckhard Kessler, Wolfenbütteler Abhandlungen zur Renaissanceforschung 21 (Wiesbaden: Harrassowitz in Kommission, 2002), 217–40.

23. Erastus, *Disputationum et epistolarum medicinalium volumen*, epistola 21 (January 26, 1576), second series of foliation, 70r–73r: "Ad D. Hieronymum Capivaccium. Quod nullus morbus, quatenus morbus, contagiosus dici possit. 2. Utrum actio aliqua laedatur, instrumento non laeso. 3. Galenici loci cap. 1 lib. 3 de sympt. caus. explicatio. 4. De putredine" (also in Crato et al., bk. 4 [1593], pp. 328–40, and Scholz, *Epistolae* [1598], epistola 270, cols. 503–8; in both undated, omitting the sentence about Monau quoted below, and with a different explicit). At 70r: "Nec minus amorem tuum erga me tum ex adolescentum, quos tibi commendaram, literis, tum ex tua ipsius epistola didici. Scribunt enim illi mihi, maxime autem Petrus Monavius, se a te non modo diligi, verum etiam vehementer amari." Peter Monau (1551–1588) studied medicine at Padua from 1575 to 1578, before obtaining his doctorate from Basel; see *ADB* 22 (1885): 163, www.deutsche-biographie.de/ sfz65048.html.

24. Scholz, *Epistolae* (1598), epistola 83 (Capodivacca to Erastus, dated 1576), cols. 121–26 (also in Crato et al., bk. 4 [1593], pp. 316–28). At col. 121: "Nunc autem quia longa oratione, miro artificio et omni diligentia meam confutare sententiam laboras de contagio, et de putredine spiritus, et de aliis, quae tibi lubenti animo per literas significavi, breviter meam mentem aperiam[.]"

25. Scholz, *Epistolae* (1598), epistola 84 (Capodivacca to Monau, November 4, 1579), cols. 126–29 (also in Crato, bk. [1] [1591], pp. 259–66); Scholz, *Epistolae* (1598), epistola 211 (Monau to Capodivacca, February 21, 1581), cols. 350–56 (also in Crato et al., bk. 5 [1593], pp. 470–83); Scholz, *Epistolae* (1598), epistola 85 (Capodivacca to Monau, n.d., but clearly a response to Monau's letter of 1581 because of the reply to his question about the current epidemic), cols. 129–31 (also in Crato, bk. [1], [1591], pp. 266–70); Scholz, *Epistolae* (1598), epistola 212 (Monau to Capodivacca, n.d.), cols. 356–58 (also in Crato et al., bk. 5 [1593], pp. 484–89).

26. Gliozzi, "Capodivacca (Capivaccio, Capivacceus), Girolamo"; Bernardino Paterno (d. 1592), *Explanationes in primam fen primi Canonis Avicennae* (Venice: apud Franciscum de Franciscis Senensem, 1596).

27. For a biography of Mercuriale (1530–1606), see Italo Paoletti, *Girolamo Mercuriale e il suo tempo* (Lanciano: Cooperativa editoriale, 1963); Alessandro Simili, "Gerolamo Mercuriale lettore e medico a Bologna: Nota 1. La condotta di Gerolamo Mercuriale a Bologna," *Rivista di storia delle scienze mediche e naturali*, ser. 6, 32 (1941): 161–96; and Giuseppe Ongaro, "Mercuriale, Girolamo," *DBI*, vol. 73 (2009), www.treccani.it/enciclopedia/girolamo-mercuriale_(Dizionario-Biografico)/. For a full bibliography of Mercuriale's printed works, see Giancarlo Cerasoli and Brunella Garavini, "La bibliografia delle opera a stampa di Girolamo Mercuriale," *Medicina & Storia* 6 (2006): 85–119. Mercuriale's principal work is now available in a critical edition: Girolamo Mercuriale, *De arte gymnastica*, ed. and trans. Concetta Pennuto, Vivian Nutton, and Jean-Michel Agasse (Florence: Leo S. Olschki, 2008). On his writings and influence, see *Girolamo Mercuriale: Medicina e cultura nell'Europa del Cinquecento; Atti del Convegno "Girolamo Mercuriale e lo spazio scientifico e culturale del Cinquecento" (Forlì, 8–11 novembre 2006)*, ed. Alessandro Arcangeli and Vivian Nutton, Bibliothèque d'Histoire des Sciences 10 (Florence: Leo S. Olschki, 2008). The Bolognese painter Lavinia Fontana is known to have painted at least four portraits of Mercuriale, which he sent to patrons and colleagues; a painting in the Walters Art Museum, Baltimore, depicting a physician seated in front of a shelf of medical books and pointing to an open copy of Vesalius's *De humani corporis fabrica libri septem* is thought to be one of them; see Caroline P. Murphy, *Lavinia Fontana: A Painter and Her Patrons in Sixteenth-Century Bologna* (New Haven: Yale University Press, 2003), 70–72, with an appendix of documents regarding the Duke of Urbino's request, and the Walters Art Museum, http://art.thewalters.org/detail/19054/portrait-of-gerolamo-mercuriale/.

28. Paul Grendler, *The Universities of the Italian Renaissance* (Baltimore: Johns Hopkins University Press, 2002), 193, and Favaro, *Atti della nazione germanica artista nello Studio di Padova*, 1:147–57.

29. Gaspard Bauhin, ed., *Gynaeciorum sive de mulierum affectibus commentarii Graecorum, Latinorum, barbarorum, jam olim et nunc recens editorum: in tres* [i.e., quatuor] *tomos digesti, et necessariis passim imaginibus illustrati*. Tomus II, *Gynaeciorum physicus et chirurgicus: continens inter caetera Hieron. Mercurialis antecessoris Patauini elegantissimi, Muliebrium libros IV. Franc. item Rousseti Hysterotomotokian e Gallico conuersam* (Basel: apud Conradum Waldkirch, 1586), "Praefatio ad lectorem." On Bauhin's career, see Gweneth Whitteridge, "Bauhin, Gaspard," in *Dictionary of Scientific Biography*, ed. Charles Coulston Gillispie, 16 vols. (New York: Charles Scribner's Sons, 1970–80), 1:522–25. Mercuriale lectured on diseases of women in 1572–73; see E. Greco, "Il posto di Girolamo Mercuriale

nella storia dell'ostetricià e ginecologia," *Rivista Italiana di Ginecologia* 45 (1961): 148–68.

30. Girolamo Mercuriale, *De morbis muliebribus praelectiones ex ore Hieronymi Mercurialis, iam dudum a' Gaspare Bauhino exceptae, ac paulo antea inscio autore edit[a]e: nunc vero per Michaelem Columbum ex collatione plurium exemplarium consensu auctoris locupletiores, & emendatiores factae* (Venice: apud Felicem Valgrisium, 1587). In another edition (Venice: apud Juntas, 1591), the preface, "Michael Columbus medicinae studiosis" (dated December, 1586), states [α3v]: "Quod sane officium [Bauhin's publication] prima facie non satis arrisit, cum antea plurimis et Venetis, et Patavinis, atque exteris typographis, illud tantopere efflagitantibus, denegasset. Postea vero ut animo est leni, summeque in Germanos benevolo, atque propenso, coepit boni aequique consulere id, quidquid fuit Germanorum in seipsum studii ac propensionis."

31. Donato Giannotti, *Lettere a Piero Vettori*, ed. Roberto Ridolfi and Cecil Roth (Florence: Vallecchi, 1932), 143, 153–54, contains references by Giannotti to Mercuriale and his correspondence with Vettori; the *carteggio* of letters addressed to Vettori included in this volume lists twenty-five letters from Mercuriale (170, 179, 182); Alessandro Simili, *Gerolamo Mercuriale lettore e medico a Bologna* (Bologna: Azzoguidi, 1966), 82–86, edits a series of letters from Mercuriale to Aldrovandi; Mercuriale's correspondence with Pinelli survives in the Biblioteca Ambrosiana (Adolfo Rivolta, *Catalogo de' codici Pinelliani dell'Ambrosiana* [Milan: Tip. Pontificia Arcivescovile S. Giuseppe, 1933]); Galileo Galilei, *Le opere di Galileo Galilei: Edizione nazionale*, vol. 10 (reprint Florence, 1934), 54–55, 74–75, 83–84, for nos. 46, 65, and 73.

32. Girolamo Mercuriale, *Responsorum, et consultationum medicinalium tomus primus. Nunc primum a Michaele Columbo collectus, et in lucem editus* (Venice: apud Iolitos, 1587), consultatio 8, pp. 25–26: "De mensibus inordinatis, atque imminutis, ac de sterilitate. Ad Abrahamum e Portalionis medicum Hebraeum." Incipit: "Quod illustrissima mulier . . ." Explicit: "Bene vale, tibique de Mercuriale ea omnia pollicearis, quae a fideli amico expectari possunt." On Abraham Portaleone and his correspondence with Christian physicians, see Andrew Berns, "The Natural Philosophy of the Biblical World: Jewish and Christian Physicians in Late Renaissance Italy" (PhD diss., University of Pennsylvania, 2011), chapter 4.

33. On the complicated publishing history of the four parts of Mercuriale's *Consultationes et responsa*, see Cerasoli and Garavina, "Bibliografia," 107–8. The first volume, edited by Michele Colombo, was published in Venice (apud Iolitos, 1587) and in Basel (apud Conradum Waldkirch, 1588); an edition of volumes 1 and 2, both edited by Colombo, followed (Venice: apud Iolitos, 1589–90). Volume 3 was published in Venice (apud Franciscum de Franciscis Senensem, 1597); volume 4, edited by Guglielmus Athenius (Mercuriale's student Willem van Thienen, from Brussels, who also wrote the volume's dedicatory letter to the papal physician

Giacomo Bonaventura), was published in Venice (apud Iuntas, 1604). The first three volumes, reedited by Mondino Mondini, were published together (Venice: apud Iacobum de Franciscis, 1619–20). All four volumes were later published together (Venice: apud Iuntas, 1624). I have consulted Girolamo Mercuriale, *Responsorum, et consultationum medicinalium tomus primus. Nunc primum a Michaele Columbo collectus, et in lucem editus* (Venice: apud Iolitos, 1587); idem, *Responsorum et consultationum medicinalium tomus quartus. Nunc primum a Guilielmo Anthenio philosopho et medico editus* (Venice: apud Iuntas, 1604); and idem, *Consultationes et responsa medicinalia quatuor tomis comprehensa. Postrema hac editione a Mundino Mundinio philosopho et medico Vincentino annotationibus exornata: Addita Mercurialis Collegiandi (ut vocant) ratione,* 4 vols. in 1 (Venice: apud Iuntas, 1624). In the copy of the last-named volume held by the library of the New York Academy of Medicine, each internal volume has a separate title page, with publisher and date, as follows: volume 1: apud Iuntas, 1624; volume 2: apud Iolitos, 1590; volume 3: apud Franciscum de Franciscis, 1597; volume 4: apud Iuntas, 1604.

34. Mercuriale, *Responsorum, et consultationum medicinalium tomus primus* (1587), consultatio 38, pp. 86–87: "De articulari dolore, et vermibus, ad Cratonem medicum clarissimum." The *consultatio* for Emperor Maximilian II is no. 111, pp. 255–61. In addition, in this volume the patients referred to in *consultationes* nos. 2, 19, 20, 31, 66, 85, 90, 97, 106, and 112 are identified as "germanus/a."

35. Had Mercuriale and Crato possibly met even earlier, during their student days in Padua? Crato had studied there under Giambattista Da Monte (d. 1551). For biographical information on Mercuriale, including his call to treat Maximilian II, see Paoletti, *Girolamo Mercuriale e il suo tempo.*

36. Mercuriale, *Consultationes et responsa medicinalia quatuor tomis comprehensa,* consultatio 15, 3:15r–16v: "De interpolata oculorum offuscatione, et variis ipsorum symptomatibus pro Illustri Imperatoris cubiculario ad Jacobum Scutellarium archiatrum"; ibid., 4:138–44: "Consultatio pro serenissimo principe Carolo Austriaco marchione Burgoviensi, sinistri lateris dolore, flatibus hypochondriacis, aliisque affectibus laborante. Paulo Weinhart medico excellentissimo, Hieronymus Mercurialis" (Charles of Austria [1540–1590] was one of the sons of Emperor Ferdinand I).

37. Mercuriale, *Responsorum, et consultationum medicinalium tomus primus* (1587), consultatio 66 (addressed to Lukas Stenglin), pp. 149–52; Mermann is the addressee for ibid., no. 106, pp. 246–47, and no. 112, pp. 262–63. On Stenglin, see Didier Kahn, *Alchimie et Paracelsisme en France à la fin de la Renaissance (1567–1625)* (Geneva: Librairie Droz, 2007), 136, 156, 186; on Mermann (1547–1625), see *ADB* 21 (1885): 447–48, www.deutsche-biographie.de/sfz62037.html. Mercuriale, *Consultationes et responsa medicinalia quatuor tomis comprehensa,* consultatio 106, 2:161–62, is also addressed to Mermann. In vol. 2 the patients referred to in consultationes 15, 18, 22, 54, and 106 are identified as "German," as are several in vol. 3.

38. Mercuriale, *Consultationes et responsa medicinalia quatuor tomis comprehensa*, consilium 10, 4:21–24: "Hieronymo Fabricio. Heronymus Mercurialis S. D. De commentariis Horatii Augenii de sanguinis missione" (on this item, see chapter 3). On Fabrizi (1533–1619), see Bruno Zanobio, "Fabrici, Girolamo," in *Dictionary of Scientific Biography*, 4:507–12. Maurizio Rippa Bonati, "Girolamo Fabrici d'Acquapendente: per una bio-crono-bibliografia," in *Il teatro dei corpi: Le pitture colorate d'anatomia di Girolamo Fabrici d'Acquapendente*, ed. Maurizio Rippa Bonati and José Pardo-Tomás (Milan: Mediamed Edizioni Scientifiche, 2004), 275–77, provides additional bibliography. Mercuriale, *Consultationes et responsa medicinalia quatuor tomis comprehensa*, consilium 44, 4:97–98: "Ad Ioannem Baptistam Codronchium medicum praestantissimum. De mucronalis chartilaginis casus superstitione." On Codronchi (1547–1628), see Carlo Colombero, "Codronchi, Giovanni Battista," *DBI*, www.treccani.it/enciclopedia/giovan-battista-codronchi_ (Dizionario-Biografico)/. Mercuriale, *Consultationes et responsa medicinalia quatuor tomis comprehensa*, 4:155–56: "Ad clarissimum virum Iacobum Bonaventuram Sanctissimi D. N. Clementis Octavi medicum." On Bonaventura (d. 1602), see Gaetano Marini, *Degli archiatri pontifici volume primo, nel quale sono i supplimenti e le correzioni all'opera del Mandosio* (Rome, 1784), 479–80.

39. Mercuriale, *Consultationes et responsa medicinalia quatuor tomis comprehensa*, consilium 44, 4:97–98: "Ad Ioannem Baptistam Codronchium medicum praestantissimum. De mucronalis chartilaginis casus superstitione." At p. 97: "Nam etiam ipse, qui et has, et alias complures simplices plebeculae nugas iamdudum compertas habeo, non semel audivi mulieres praesertim animam cadutam nominantes, quasi vero chartilago illa, mucronata ab anatomicis appellata, vi daemonum, aut effascinationum, ita a suo loco supra stomachum cadat, ut omnes illius functiones deperdat, nisi verbis, et manibus in suam sedem restituatur. Qua in re multos errores ob ignorantiam committi non dubito, quorum ille prior est de ipsius vocis abusu, quandoquidem ita partem illam vocarunt, tamquam sit quoddam animae domicilium, quo descendente simul anima ipsa cadat, et eo restituo simul anima ad propriam sedem redeat, quo figmento nil absurdius, nil ineptius reperiri possit."

40. Ibid., 4:155–56: "Ad clarissimum virum Iacobum Bonaventuram Sanctissimi D. N. Clementis Octavi medicum." Incipit: "Quinquaginta prope anni aguntur, in quibus medicinam exercere, atque public docere Romae, Patavii, Bononiae, ac Pisis non destiti, quo tempore etiam a Maximiliano secundo Imperatore pro ipsius valetudine vocatus, aliisque maximis Principibus, et viris, tum coram, tum per literas inservire compulsus, innumeras fere, et scribendi tam Italice, quam Latine, et medendi occasiones nactus sum, ut propterea mea responsa medicinae, quae vulgo consilia nuncupantur, quasi fato quodam sint, et exscripta, et publicata, cumque in omnes pene totius Europae regiones pervenerint, aliquos fuisse intellexi, qui summis laudibus ea susceperunt, aliquos vero qui longiores discur-

sus, maioremque, et doctrinae, et remediorum ubertatem optaverint, quorum omnium dissidia non arduum esset componere . . . idque potissimum in aegrotantibus
consilia expostulantibus invenire certum est, qui cum nihil aliud a medicis con-
sultutis quaerant, quam ut aegritudinem suarum naturam explicent, curandique
rationem edoceant, haudquaquam profusas, et inutiles disputationes expetunt, sed
duntaxat ut sibi compendio quodam res suae aperiantur, remediaque opportuna
monstrentur. Quod etiam Cornelius Celsus intelligere voluit, ubi aegros non verbis,
et eloquentia sed remediis curari scripsit . . . Nam quicunque varia me commentaria
perlegerunt, quique ex suggestu in diversis Italiae gymnasiis me publice docentem
auscultaverunt, facile, et de morbis, de rerum natura, aliis que similibus dissere non
difficulter posse cognoverunt."

41. Mercuriale's letters to Zwinger are found in Basel, Universitätsbibliothek,
in the Frey-Grynaeus collection (mostly in the series Frey-Gryn. II.4, II.19, and
II.23), and listed in the library's online *Katalog Handschriften und Nachlässe*. Mercuriale's last letter to Zwinger (Frey-Gryn. II.19.70) was written in July 1588;
Zwinger died in March 1588. On these letters, see Antonio Rotondò, *Studi e ricerche di storia ereticale italiana del Cinquecento* (Turin: Edizioni Giappichelli,
1974), 287, 399–407, 546–48, with complete or partial editions of seven of Mercuriale's letters to Zwinger; and, on Mercuriale's letters to Crato, ibid., 400n10 (Rotondò points out the avoidance of anything religiously controversial); idem, "La
censura ecclesiastica e la cultura," in *Storia d'Italia*, vol. 5, *I documenti* (Turin:
Giulio Einaudi editore, 1973), 1399–1492, at 1449–51. Mercuriale's correspondence
with Zwinger is also briefly discussed in Paul F. Grendler, *The Roman Inquisition
and the Venetian Press, 1540–1605* (Princeton: Princeton University Press, 1973)
186–87. For further discussion, see Nancy G. Siraisi, "Mercuriale's Letters to
Zwinger and Humanist Medicine," in Arcangeli and Nutton, *Girolamo Mercuriale*, 77–95.

42. Mercuriale owned almost 1,200 volumes in 1587; see Jean Michel Agasse,
"La bibliothèque d'un médecin humaniste: L'index librorum de Girolamo Mercu-
riale," *Les Cahiers de l'Humanisme* 3–4 (2002–03): 201–53, with the inventory of
Mercuriale's books at 215–53.

43. Girolamo Mercuriale, *Liber responsorum et consultationum medicinalium
nunc primum a Michaele Columbo collectus et in lucem editus* (Basel: Per Conradum
Waldkirch 1588), which I cite from the online catalogue of the National Library of
Medicine. For Mercuriale's letter to Zwinger enclosing the work, see Basel, Universitätsbibliothek, Frey-Gryn. II.4.208 (August 23, 1587).

44. Basel, Universitätsbibliothek, Frey-Gryn. II.4.187 (June 26, 1579): "relatum
est mihi Erastum virum doctissimum meique amicissimum brevi Basileam ventu-
rum . . . id ut mihi significes utrum verum sit vehementer peto, cum nam ipsius
eruditione atque literis maximopere delecter non possum non summopere gaudere, de illius ad vos adventu, a quo loco facilius erit et scribere et literas accipere."

Mercuriale inquired after or sent greetings to Erastus in several other letters to Zwinger.

45. On Dudith, see Pierre Costil, *André Dudith, humaniste hongrois, 1533–1589: Sa vie, son oeuvre et ses manuscrits grecs* (Paris: Société d'édition "Les Belles lettres," 1935). Dudith wrote numerous letters to Raphanus on medical topics; see, for example, Scholz, *Epistolae* (1598), epistolae 16–34 (all written in the period 1577–83), cols. 20–46, and Costil, *André Dudith*, 9–10.

46. Scholz, *Epistolae* (1598), epistola 92 (Mercuriale to Raphanus, 1571), cols. 142–45 (also in Crato et al., bk. 2 [1592], pp. 261–66). At col. 143: "De philtris, quae ad conciliandum amorem mirifice conferre antiquitas credidit, mea sententia est, neque illa existere, nec medicorum prudentione contemplatione subiici . . . Similitudo morum, studiorum, aetatis est potissimum amoris philtrum, et etiam actiones et vita honeste composita." Ibid., epistola 93 (Mercuriale to Raphanus, May 16, 1571), col. 145 (also in Crato et al., bk. 2 [1592], pp. 266–68).

47. Scholz, *Epistolae* (1598), epistola 30 (Dudith to Raphanus, August 21, 1582), cols. 38–39: "Memini me aliquando Mercurialis, nisi fallor, ea de re [in margin: philtra] ad te epistolam legere: sed excidit e memoria. Dum tu mihi, per ocium, aliquid de iis rebus scribis, rogo te, mone me de autore aliquo, in quo haec pertractentur."

48. Scholz, *Epistolae* (1598), epistola 88 (Mercuriale to Dodoens, Padua, June 21, 1575), cols. 132–34 (also in Crato et al., bk. 4 [1593], pp. 376–80).

49. Scholz, *Epistolae* (1598), epistola 86 (Mercuriale to Crato, Padua, 1580), cols. 131–32 (also in Crato et al., bk. 2 [1592], pp. 232–33); Scholz, *Epistolae* (1598), epistola 87 (Mercuriale to Crato, Padua, September 22, 1580), col. 132 (also in Crato et al., bk. 2 [1592], pp. 234–35); Scholz, *Epistolae* (1598), epistola 105 (Crato to Mercuriale, October 28, 1580), cols. 183–88 (also in Crato et al., bk. 2 [1592], pp. 235–46); Scholz, *Epistolae* (1598), epistola 106 (Crato to Mercuriale, n.d.), cols. 188–89 (also in Crato et al., bk. 2 [1592], pp. 246–48).

50. Nutton, "Reception of Fracastoro's Theory of Contagion."

51. On Monau's appointment, see R. J. W. Evans, *Rudolf II and His World: A Study in Intellectual History, 1576–1612* (New York: Oxford University Press, 1984), 203.

52. Scholz, *Epistolae* (1598), epistola 213 (Monau to Mercuriale, Prague, December 24, 1581), cols. 358–60 (also in Crato et al., bk. 5 [1593], pp. 489–93). See Galen, *De simplicium medicamentorum temperamentis et facultatibus* 9.23, in his *Opera omnia*, ed. C. G. Kuhn, 20 vols. (Leipzig, 1822–30), 12:230.

53. Scholz, *Epistolae* (1598), epistola 89 (Mercuriale to Monau, Padua, February 17, 1581), cols. 134–36 (also in Crato, bk. [1] [1591], pp. 270–74). At col. 135: "Haec ignorasse gravissimum scriptorem Galenum valde miror, aut saltem dissimulare, cum scripsit, esse in plumbo multam substantiam humidam congelatam aeream, et paucam terream. Sed multo magis ipsius simplicitatem admiror, qui dixit, cavernosum esse plumbum, ac propterea pondus eius, ut statuarum vincula probant, diuturnitate augeri, quod aereum sit." Compare Pietro Andrea Mattioli, *Commen-*

tarii in libros sex Pedacii Dioscoridis Anazarbei, De medica materia (Venice: in officina Erasmiana, apud Vincentium Valgrisium, 1554), 5.58, p. 595: "Quod vero et aereae sit particeps [plumbum], hoc habeto signum omnium, quae novimus, unicum plumbum tum mole ipsa, tum pondere augetur, si condatur in aedibus subterraneis, aerem habentibus turbidum, ita ut quaecunque illic ponantur, celeriter situm colligant. Tum etiam plumbea statuarum vincula, quibus earum pedes annectuntur, saepenumero crevisse visum est, et quaedam adeo intumuisse, ut ex lapidibus dependerent crystalli modo."

54. Scholz, *Epistolae* (1598), epistola 214 (Monau to Mercuriale, Vienna, "ipso die Pentecostes secundum veteres fastos" 1583 (i.e., according to the Julian calendar, rather than the recently introduced Gregorian calendar), cols. 360–65 (also in Crato et al., bk. 5 [1593], pp. 493–506). At col. 362: "Sicuti enim philosophis cum medicis belle convenit, quantumvis illi corpus humanum ex quatuor elementis, igne, aere, aqua et terra. Hi vero ex quatuor humoribus, sanguine, phlegmate, et bile utraque constare decernant: ita etiam non pugnant inter se Peripateticos, et quos chymicos, veluti in fundendorum metallorum arte ac peritia eximie versatos, recentiori vocabulo appellant, etiamsi illi una cum Aristotele, quem unicum ducem sequuntur, metalla et fossilia a duplici exhalatione vaporosa et fumosa ortum trahere dicunt. Hi autem geminum illis principium, quae sulfur et mercurium indigerant, assignant. Illi enim in remotioribus tantummodo causis, quas sola mentis acies apprehendit, demonstrandis occupantur." At col. 363–64: "Quid igitur mirum obsecro, aut supra fidem communem est, si tam ignobile et imperfectum metallum, quod vitam nunquam habuit, nec habere aptum est, quodque minime exquisita aut opera transumptione, vel praeparatione materiae indiget, credamus aliquod augmentum extrinsecus accipere posse, si et idonea materia, aer videlicet crassus et turbidus, qualis in conclusis locis est, subministretur, agentique praeterea causa vere proportionata non destituatur."

55. Scholz, *Epistolae* (1598), epistola 214 (Monau to Mercuriale, Vienna, Pentecost 1583), col. 365.

56. Ibid., epistola 90 (Mercuriale to Monau, Padua, December 20, 1583), cols. 136–38 (also in Crato, bk. [1] [1591], pp. 274–78). Incipit: "Optassem Monavi mi, aut plus otii mihi superesse, aut curarum minus habere, quo praeclara tuae pro Galeni circa plumbi naturam suscepta diffensione disceptationi responderem: sed quia ea est omni tempore, et praesertim haec nostri ordinis hominum conditio, ut vix tundendis unguibus prae infirmorum visitationibus, publicis lectionibus, atque studiis facultas concedatur."

57. Scholz, *Epistolae* (1598), epistola 215 (Monau to Mercuriale, Prague, January 13, 1584), cols. 365–71 (also in Crato et al., bk. 5 [1593], pp. 506–21, where it is dated January 31), with comments on Mercuriale, *De morbis puerorum tractatus . . . ,* 1st ed. (Venice: apud Paulum Meietum Bibliopolam Pat., 1583), at cols. 366–67, and on Mercuriale, *De venenis et morbis venenosis tractatus . . . ex voce excellentissimi Hieronymi Mercurialis Forliviensis medici clarissimi diligenter excepti et*

in libros duos digesti: Opera Alberti Scheligii, 1st ed. (Venice: apud Paulum Meietum Bibliopolam Pat., 1584) (in which Schelig's preface to the King of Poland is
dated March 1, 1583), at cols. 367–68. Monau evidently had a copy of the latter
work in his hand by January 1584. Possibly he had received—perhaps from the
Polish student at Padua who edited the work for publication—a proof or prepub-
lication copy of the book.

    58. Scholz, *Epistolae* (1598), epistola 216 (Monau to Mercuriale, Prague, Febru-
ary 7, 1584), cols. 372–74 (also in Crato et al., bk. 5 [1593], pp. 521–26).

    59. Scholz, *Epistolae* (1598), epistola 91 (Mercuriale to Monau, Padua, March 31
1584), cols. 138–42 (also in Crato, bk. [1] [1591], pp. 279–89).

    60. Charles D. Gunnoe and Jole Shackelford, "Johannes Crato von Kraff-
theim (1519–1585): Imperial Physician, Irenicist, and Anti-Paracelsian," in *Ideas
and Cultural Margins in Early Modern Germany: Essays in Honor of H. C. Erik
Midelfort,* ed. Marjorie Elizabeth Plummer and Robin B. Barnes (Farnham, Surrey: Ashgate, 2009), 201–16. It is less clear to what extent Peter Monau can be regarded as anti-Paracelsian. Between 1580 and 1582 Monau received a number of
letters from the English Paracelsian Thomas Moffett; some of them discuss chymical remedies, though without any explicit reference to Paracelsus (Scholz, *Epistolae* [1598], epistolae 277–80, cols. 529–33).

    61. See Charles Webster, "Conrad Gessner and the Infidelity of Paracelsus," in
*New Perspectives on Renaissance Thought: Essays in the History of Science, Education
and Philosophy in Memory of Charles B. Schmitt,* ed. John Henry and Sarah Hutton
(London: Gerald Duckworth, 1990), 13–23, and Dane T. Daniel, "Coping with
Heresy: Suchten, Toxites, and the Early Reception of Paracelsus's Theology," in
*Chymists and Chymistry: Studies in the History of Alchemy and Early Modern Chem-
istry,* ed. Lawrence M. Principe (Sagamore Beach, MA: Chemical Heritage Foun-
dation and Science History Publications, 2007), 53–62.

    62. Mercuriale to Zwinger, Basel, Universitätsbibliothek, Frey-Gryn. II.4.164
(November 21, 1573): "Expectabo igitur magna cupiditate . . . Erasti anti Thessali
quartum tomum." The work referred to is presumably part 4 of Thomas Erastus,
*Disputationum de medicina nova Philippi Paracelsi pars prima [–quarta] . . . acces-
sit Tractatio de causa continente eodem authore* (Basel: apud Petrum Pernam,
[1571?]–72). On Zwinger and Paracelsus, see Carlos Gilly, "Zwischen Erfahrung
und Spekulation: Theodor Zwinger und die religiöse und kulturelle Krise seiner
Zeit," *Basler Zeitschrift für Geschichte und Altertumskunde* 77 (1977): 57–137; I
am grateful to Ann Blair for calling my attention to this article. For the works
of Paracelsus and Erastus owned by Mercuriale, see Agasse, "La bibliothèque
d'un médecin humaniste." Erastus, *Disputationum et epistolarum medicinalium volumen,* epistola 17 (to Capodivacca), second series of foliation, at 50v: "in
tercia Disputationum nostrarum adverso Paracelsum parte, valide ac dilucide
probavi."

63. See Antonio Clericuzio, "Chemical Medicine and Paracelsianism in Italy," in *The Practice of Reform in Health, Medicine, and Science, 1500–2000: Essays for Charles Webster*, ed. Margaret Pelling and Scott Mandelbrote (Aldershot, UK: Ashgate, 2005), 59–79, and Richard Palmer, "Pharmacy in the Republic of Venice," in *The Medical Renaissance of the Sixteenth Century*, ed. Andrew Wear, R. K. French, and I. M. Lonie (Cambridge: Cambridge University Press, 1985), 100–117, at 111–14.

64. Scholz, *Epistolae* (1598), epistola 98 (Donzellini to Crato, March 10, 1585), cols. 155–56 (also in Crato et al., bk. 6 [1611], epistola 40, pp. 112–15). At col. 155: "De chymicis tecum sentio, nec ego illis utor, nisi in chronicis quibusdam de quibus meam sententiam intelliges brevi, cum leges libellum quendam a me nuper editum contra quendam Empiricum chymicum, qui librum edidit, cui titulum fecit Il flagello di Medici rationali, Italica ista sunt. Ego alieno nomine ita illum tractavi, ut illum suae audaciae poenituerit . . . Ego aliquando libros omnes Paracelsi apud me habui; sed cum ex eorum lectione nihil proficerem, pro aliis permutavi, nec in mea Bibliotheca tales libros haerere volui, qui nihil docerent."

65. See Tommaso Zefiriele Bovio, *Flagello de' medici rationali . . .* , in his *Opera contra medici putaticii rationali* (Padua: Pietro Paolo Tozzi, 1626; the first edition of Bovio's *Flagello* appeared in 1583). On Bovio, see Mariacarla Gadebusch Bondio, "Paracelsismus, Astrologie und ärztliches Ethos in den Streitschriften von Tommaso Bovio (1521–1609)," *Medizinhistorisches Journal* 38 (2003): 215–44, with a discussion of Bovio's relation to Donzellini, Donzellini's response (published under the name of Claudio Gelli) to Bovio's treatise, and Donzellini's letter to Crato at 237–41.

66. See Palmer, "Physicians and the Inquisition in Sixteenth-Century Venice."

67. Ibid.

68. Martin, *Venice's Hidden Enemies*, 220–24.

Chapter 2 · *The Court Physician Johann Lange and His* Epistolae Medicinales

1. Nicolas Reusner, "Praefatio," in Johannes Lange, *Epistolarum medicinalium volumen tripartitum, denuo recognitum, et dimidia sui parte auctum* (Frankfurt: apud heredes Andreae Wecheli, Claudium Marnium & Ioannem Aubrium, 1589; cited hereafter as Lange, *Epistolae* [1589]): "Ad clarissimum excellentissimumque virum D. Georgium Wirth, philosophiae et medicinae doctorem, olim Caroli V Aug. Caes. et Philippi II regis Hispaniarum medicum, cognatum suum carissimum Praefatio Nicolai Reusneri Iurisconsulti." At [α8v]–βr: "Quorum in numero sunt Miscellanea ista *Epistolarum medicinalium*, varia ac rara eruditione, tum etiam rerum omnis generis scitu dignissimarum explicatione referta . . . Sane non ignoro, complures in hoc scriptorum genere praeclaram posuisse operam: similibus Epistolis Medicinalibus conscribendis, atque publicandis: in quibus prae-

cipuam laudem et commendationem merentur Ioannes Manardus, Aloysius Mundella, Ioannes Baptista Theodosius, Petrus Andreas Mathiolus, Nicolaus Massa, Victor Trincavellius, et si qui praeter hos nominatos sunt alii, quorum nomina comperta non habeo. Sed ex Germanis (nam hi, quos nominavi, omnes Itali sunt) solus, nisi fallor, Langius hoc scriptionis genus felicissime tractavit." The preface, dated 1589, also appears in Johannes Lange, *Epistolarum medicinalium volume tripartitum* . . . (Hanau: typis Wechelianis, apud Claudium Marnium & haeredes Ioannis Aubrii, 1605; cited hereafter as Lange, *Epistolae* [1605]).

2. Reusner, "Praefatio," in Lange, *Epistolae* (1589), α3v: "Tum vero argumento huius rei esse potest perpetua artis illius [medicinae] possessio in familiis praecipuis: ut quod de Asclepiadarum familia proditum est . . . Quod et in aliis artibus ac disciplinis observari ita solitum fuit."

3. For general accounts of Lange and his *epistolae medicinales* by historians of medicine over the past century, see Viktor Fossel, "Aus den medizinischen Briefen des pfalzgräflichen Leibarztes Johannes Lange," *Archiv für Geschichte der Medizin* 7 (1913): 238–52; Ralph H. Major, "Johannes Lange of Heidelberg," *Annals of Medical History*, n.s., 7 (1935): 133–40; and, especially, Vivian Nutton, "John Caius und Johannes Lange: medizinischer Humanismus zur Zeit Vesals," *NTM: Zeitschrift für Geschichte der Wissenschaften, Technik und Medizin* 21 (1984): 81–87, and idem, "Humanist Surgery," in *The Medical Renaissance of the Sixteenth Century*, ed. Andrew Wear et al. (Cambridge: Cambridge University Press, 1985), 75–99, especially 91–96.

4. Biographical summary in *ADB* 17 (1883): 637–38, www.deutsche-biographie .de/sfz47933.html. For near-contemporary biographical accounts of Lange, see Melchior Adam, *Vitae Germanorum medicorum: qui seculo superiori, et quod excurrit, claruerunt. Congestae et ad annum usque MDCXX deductae* (Heidelberg, 1620), pp. 140–54[misprint for 144]; Nicolas Reusner, *Icones seu imagines virorum literis illustrium qui seculo XV praesertim doctrina Religionis aliarumque bonarum scientia tamquam lumina in nostra Germania claruere* (Frankfurt, 1719; 1st edition 1587), 88–89; and Reusner, "Praefatio," in Lange, *Epistolae* (1589), [α7v–α8r].

5. Jürgen Leonhardt, "Classics as Textbooks: A Study of the Humanist Lectures on Cicero at the University of Leipzig, ca. 1515," in *Scholarly Knowledge: Textbooks in Early Modern Europe*, ed. Emidio Campi, Simone De Angelis, Anja-Silvia Goeing, and Anthony Grafton (Geneva: Librairie Droz, 2008), 89–112, at 98–105; Jill Kraye, "Aristotle's God and the Authenticity of *De mundo*: An Early Modern Controversy," *Journal of the History of Philosophy* 28 (1990): 339–58, at 355–56. Kraye notes (355n74) an edition of Lange's *De mundo* commentary published in Frankfurt, 1606.

6. Lange, *Epistolae* (1605), 1.35, at p. 150: "dum in Lipsensi academia, frequenti discipulorum consessu Procli sphaeram, et reliqua cosmographiae rudimenta interpretabar"; ibid., 1.63, pp. 304–16: "De novis Americi orbis insulis, antea ab Hannone Carthaginense repertis: et geographiae utilitate."

7. Johannes Lange, *Oratio Ioannis Langij Lembergij: Encomium theologicae disputationis Doctorum Ioannis Eckij, Andreae Carolostadij, ac Martini Lutherij complectens. Illustriss. Principi D. ac D. Georgio Saxoniae duci &c, dicata. & illius iussu, cum gratiarum actione, xvi. Iulij die recitata, in frequentissima summorum uirorum concione* [colophon: Leipzig: apud Melchiorem Lottherum, 1519].

8. Nutton, "Humanist Surgery," 93.

9. Ludovicus de Leonibus, in Lange, *Epistolae* (1605), 1.47, pp. 208–11, with a full description of one of Ludovicus de Leonibus's cases; on Ludovicus de Leonibus, see David Lines, "Natural Philosophy in Renaissance Italy: The University of Bologna and the Beginnings of Specialization," *Early Science and Medicine* 6 (2001): 267–323, at 301–2, and Andrea Cristiani, ed., *I lettori di Medicina allo Studio di Bologna nei secoli XV e XVI* (Bologna: Edizioni analisi, 1988), no. 203, p. 28 (my thanks to David Lines for information regarding Ludovicus de Leonibus). For Lange's visit to the aged Leoniceno, see Lange, *Epistolae* (1605), 2.2, p. 472; the reference to Pomponazzi is found in ibid., 2.34, p. 647.

10. Lange, *Epistolae* (1605), 2.2, pp. 471–97. At pp. 472–73: "Petrus Aegineta, ex Aegina Graecorum in Peloponneso insula natus, Pauli Aeginetae medici vernaculus: qui olim Bononiae lepidissimas Aristophanis comoedias praelegerat, ac me et Leonem Pontificem Graecae linguae rudimenta docuerat." Pietro of Egina taught Greek at the University of Bologna, 1510–27; see Luigi Simeoni, *Storia della Università di Bologna*, vol. 2, *L'età moderna* (Bologna: Zanichelli, 1947), 46.

11. On his trips to Padua and Venice, see Lange, *Epistolae* (1605), 1.59 (student vacation), pp. 281–87; 1.71, at p. 379 (student visit to Pico).

12. Adolf Hasenclever, ed., "Die tagebuchartigen Aufzeichnungen des pfälzischen Hofarztes Dr. Johannes Lange über seine Reise nach Granada im Jahre 1526," *Archiv für Kulturgeschichte* 5 (1907): 385–439. Friedrich (1482–1556; Prince-Elector of the Palatinate, 1544–56) served as a general in the imperial army for part of his career. Regarding Lange's return to northeastern Italy in the 1540s, see Lange, *Epistolae* (1605), 1.64–68, pp. 316–68; ibid., 1.71, at p. 378; and Nutton, "Humanist Surgery," 302n68. However, it is unclear to what extent the travels and conversations recorded in these letters are literary creations.

13. Johannes Lange, *Medicum de republica symposium* ([n.p.], 1554; cited hereafter as Lange, *Symposium* [1554]); idem, *Medicinalium epistolarum miscellanea . . .* (Basel: Per Joannem Oporinum [1554]; cited hereafter as *Miscellanea* [1554]); idem, *Themata aliquot chirurgica, ex opere Epistolarum ipsius medicinalium*, in *Chirurgia: De chirurgia scriptores optimi quique veteres et recentiores, plerique in Germania antehac non editi, nunc primum in unum conjuncti volumen . . .* (Zurich: Per Andream Gesnerum f. et Jacobum Gesnerum fratres, 1555), 311r–320r (=Lange, *Miscellanea* [1554], items 3–10, 32, 77, 82); idem, *Medicinalium epistolarum miscellanea* (=Lange, *Miscellanea* [1554]), in *Epistolae medicinales diversorum authorum* (1556), pp. 474–557; idem, *Secunda epistolarum medicinalium miscellanea, rara et varia eruditione*

referta, non medicinae modo, sed cunctis naturalis historiae studiosis plurimum profu-
tura (Basel[: ex officina Nicolai Brylingeri, expensis Ioann. Oporini], 1560; cited
hereafter as *Miscellanea* [1560]) (this volume also includes an index to *Prima miscel-*
lanea [i.e., Lange, *Miscellanea* 1554]); idem, *Epistolae* (1589); idem, *Epistolae* (1605).

Later publications of individual letters include 2.13 and 2.14 (on the question
of "new" diseases) with Baudouin Ronsse, *De magnis Hippocratis lienibus . . . seu*
vulgo dicto scorbuto commentariolus. Accessere . . . Ioannis Langii epistolae duae de
scorbuto (Wittenberg: Clemens Schleich excudebat, 1585), pp. 263–78; 2.29 (on the
benefits of eating cheese) and an excerpt from 2.48 (on purging, especially the
ancient Egyptian use of emetics) with Diocles Carystius, *Aurea ad Antigonum Re-*
gem epistola, De morborum praesagiis, et eorumdem extemporaneis remediis. Ad haec
Arnaldi a Villanova . . . Consilium ad Regem Aragonum, de salubri hortensium usu,
ed. A. Mizauld (Paris: apud Federicum Morellum Regium Typographum, 1572), at
26r–27v and 7v–8v, respectively; an excerpt from 2.45 (on sleepwalking) in Jacob
Horst, *De aureo dente maxillari pueri Silesii, primum, utrum eius generatio natu-*
ralis fuerit necne; deinde an digna eius interpretatio dari queat. Et De noctambulo-
num natura, differentiis et causis, eorumque tam praeservativa quam etiam curativa
denuo auctus liber (Leipzig: impensis Valentini Voegelini Bibliop., 1595). The treatise on sleepwalking occupies pp. [159]-256, and it is followed by a collection of
excerpts: "De noctambulonibus omnia quae extant apud omnes medicos atque
philosophos tam veteres, quam recentiores"; the authors excerpted in this section
are Hippocrates, Aristotle, Galen, Coelius Rhodiginus, Fracastoro, Levinus Lem-
nius, Petrus Salius Diversus, Antoine Mizauld, and Lange.

I note the following later publications of individual letters from the online cata-
logue of the National Library of Medicine: 1.1 (faults of modern practitioners) with
F. Tidicaeus, *In iatromastigas, de recto et salutari usu . . . artis medicinae libellus . . .*
(Torunii Borussorum, 1592); 1.11 and 1.83 (against uroscopy) with P. van Foreest, *De*
incerto fallaci urinarum iudicio (Leiden, 1589); 1.18 and 2.23 (plague treatments) in
Joachim Baudiss, *General . . . Ordnung oder methodus wieder diess pestilentialische*
Feber ([Neyss], 1567). Letters 2.13 and 2.14 also appeared with D. Sennert et al.,
De scorbuto tractatus (Wittenberg, 1624). Letters 2.29 and 2.48 also appeared in a
reissue of the volume *Aurea ad Antigonum Regem epistola . . .* , edited by Mizauld
(Paris, 1573), as well as in *Historia hortensium* (Cologne, 1576) and *Jardinage* ([Lyon],
1578, 1605; a French version of *Historia hortensium*).

In what follows, for ease of reference I cite the letters (including those that are
general essays or incorporate a dialogue) by book, letter (epistola), and page number from the 1605 edition of Lange's *Epistolae* (in three books), with page numbers
in the original independent editions of books 1 and 2—cited as *Miscellanea* [1554]
and *Miscellanea* [1560], respectively—added in parentheses. I cite Lange's *Sympo-*
sium only in the original edition of 1554, which contains a colophon establishing
the date of composition as 1547 (the reprints in Book 3 of the 1589 and 1605 editions of his *Epistolae* lack both the colophon and an introductory poem).

14. Lange, *Epistolae* (1605), preface: "Langius candido Lectori suam scribendi epistolas medicinales occasionem recenset." At p. 1: "Elapsis iam plus minus triginta, candide lector, annis quum relictis cultoribus Enyclopaediae Musis, quas tum Lipsiae profitebar, animum ad paulo severius medicinae studium appulissem, et ne (quod Seneca inertis esse ingenii testatur) relictis fontibus rivulos sectari viderer, dum divini Hippocratis, ac fidi eius interpretis Galeni, opera diurna nocturnaque lucubratione perlustrarem[.]" Lange left his position teaching arts at Leipzig soon after 1519.

15. See, for example, Lange, *Epistolae* (1605), 2.23, pp. 590–96 (= *Miscellanea* [1560], pp. 99–103).

16. Lange, *Epistolae* (1605), 1.41, pp. 176–80 (= *Miscellanea* [1554], pp. 145–49).

17. For example, Lange, *Epistolae* (1605), 1.32 (on fallacious treatments of erysipelas), at p. 132 (= *Miscellanea* [1554], p. 109): "Hippocrates, Galenus, Avicenna, Paulus et Aetius, medicorum coryphaei, erysipelas curaturi post vacuationes rite factas, fomentationes ex succis vel decocto solatri, lactucae, sempervivi, umbilico Veneris, aut emplastra . . . applicuerunt." Scholastics named by Lange include Turisanus (Torrigiano de' Torrigiani, d. ca. 1320), cited as "Drusianus," Lange, *Epistolae* (1605), 2.18, p. 568 (= *Miscellanea* [1560], p. 81); as well as Pietro d'Abano (d. 1315) and Gentile da Foligno (d. 1348), idem, *Epistolae* (1605), 1.66, at pp. 332–33 (= *Miscellanea* [1554], p. 274), although Pietro and Gentile are mentioned with disapproval ("qui nimium arguti, plus pro versatilis ingenii ostentatione, quam pro veritate, depugnant").

18. Concerning the errors of surgeons, see Lange, *Epistolae* (1605), 1.1, pp. 8–11 (= *Miscellanea* [1554], pp. 7–10); idem, *Epistolae* (1605), 1.3–10, pp. 15–43 (= *Miscellanea* [1554], pp. 12–36). Also see Nutton, "Humanist Surgery," for a discussion of this material. Denunciations of uroscopy are found in Lange, *Epistolae* (1605), 1.11, pp. 44–49, and 1.83, pp. 459–66 (= *Miscellanea* [1554], pp. 36–41 and 377–83); idem, *Epistolae* (1605), 2.23, pp. 590–96, and 2.40–41, pp. 673–82 (= *Miscellanea* [1560], pp. 99–103 and 165–72).

19. Lange, *Epistolae* (1605), 1.35, pp. 150–60 (= *Miscellanea* [1554], pp. 124–33); idem, *Epistolae* (1605), 1.36, pp. 160–63 (= *Miscellanea* [1554], pp. 133–35). For other denunciations of divinatory astrology in Lange's letters, see also idem, *Epistolae* (1605), 1.53, pp. 243–50 (= *Miscellanea* [1554], pp. 199–206), and idem, *Epistolae* (1605), 2.60, pp. 764–75 (= *Miscellanea* [1560], pp. 239–46). On Manardo's attitude, which reflected the influence of Giovanni Pico, see Paola Zambelli, "Giovanni Mainardi e la polemica sull'astrologia," in *L'opera e il pensiero di Giovanni Pico della Mirandola nella storia dell'umanesimo: Atti del convegno internazionale, Mirandola (15–18 settembre 1963)*, 2 vols. (Florence: L. S. Olschki, 1965), 2:205–79.

20. Lange, *Epistolae* (1605), 1.53, pp. 243–50 (= *Miscellanea* [1554], pp. 199–206): "De origine alchimiae, distillationis aquarum, et olei modis, ac illorum usu." At pp. 243–45: "Deinde quod vulgus, sollicitus futurorum explorator, et percunctator,

ac avidus lucri auceps, eas acceptat artes, qua tam infirmis fulciuntur fundamentis, ut nihil duraturum diu superstrui possit: quales sunt vatidica illa astrologia, et spe dives alchimia. Quorum prima rudimenta, Aegyptii primi omnium
artium auctores excogitarunt: ad quos ob id Pythagoras et Plato navigarunt, et
quibus non inepto aenigmate Moses Aegyptum egressurus, populum Israel aurea
et argentea vasa, id est reconditos scientiarum thesaurus surripere iussit . . . Cog-
ita, obsecro, quanto vel hodie sublimationis studio et impendio, ex optimo vino et
omnibus terrae nascentibus, quintam essentiam, quam suum appellant coelum,
Mercuriumque vegetabilem, elicere per alembicum laborant: quam tandem, ut
lento putredinis calore defaecatior reddatur, intorto vitrei vasis involucro, aut pel-
licano Hermetis vase exceptum, in acinorum acervos, vel olearum recrementa, aut
si diis placet, in fimo diutius quam mensem recondunt. Tali quidem chimistarum
artificio pleraque pharmaca, aquas salutares et causticas, ex aromatibus olea odor-
ifera, leniterque edaces ex hydrargyro pulveres, putridi ulceris sordes abstergentes,
quibus momento auri addito, temere pestis virus expiare audent, et pluraque alia
magni in cura morborum usus, confici posse non abnego." But Lange concluded
(at p. 250): "cuius artis perfectionem si quaesiveris, eius ministri ut plurimum fugi-
tivi, pro carbonibus cineres, pro metallis et auro cinerum favillas micantes, tibi in
furno relinquent. Vale et cave."

21. Lange, *Epistolae* (1605), 1.79, at pp. 435–36: "te fideliter admoneo, ne agyrtis
illis medicis confidas, qui suo auro potabili et elixire indefectam iuventutem, et (ut
ille Aegyptiorum philosophus) immortalitatem promittunt."

22. The large number of sixteenth-century German medical manuscripts,
most of them collections of recipes, held by the Heidelberg University Library is
striking testimony to this vernacular medical culture; see *Kataloge der Universitätsbibliothek Heidelberg*, vol. *7, Die Codices Palatini germanici in der Universitätsbibliothek Heidelberg (Cod. Pal. germ. 182–303)*, ed. Matthias Miller and Karin
Zimmermann (Wiesbaden: Harrassowitz Verlag, 2005), http://archiv.ub.uni-heidel
berg.de/volltextserver/frontdoor.php?source_opus=8469, and the same editors' *Die
medizinischen Handschriften unter den Codices Palatini germanici der Universitätsbibliothek Heidelberg* (Heidelberg, 2005), http://archiv.ub.uni-heidelberg.de/voll
textserver/frontdoor.php?source_opus=5709. On the concept of a medical marketplace in early modern Europe, see Harold J. Cook, *The Decline of the Old Medical
Regime in Stuart London* (Ithaca: Cornell University Press, 1986), which initiated
the discussion; for an analysis and critique of recent historiography, see Mark S. R.
Jenner and Patrick Wallis, "The Medical Market Place," in *Medicine and the Market in England and Its Colonies, c. 1450–c. 1850*, ed. Mark S. R. Jenner and Patrick
Wallis (Basingstoke, UK: Palgrave Macmillan, 2007), 1–23.

23. For example, in Thomas Erastus, *Disputationum et epistolarum medicina-
lium volumen doctissimum. Nunc recens in lucem editum, opera et studio Theophili
Maderi philosophiae ac medicinae doctoris ac physices in Academia Heydelbergensi
Professoris ordinarii* (Zurich: apud Ioannem Wolphium, typis Frosch., 1595), a

number of Erastus's letters are dated, as are many of the letters in Orazio Augenio, *Epistolarum et consultationum medicinalium prioris tomi libri XII* (Venice: apud Damianum Zenarium, 1602), and all the letters in Luigi Mondella, *Epistolae medicinales, nunc ab ipso autore auctae et recognitae: in quibus variae & difficiles quaestiones utiliter tractantur: Galeni, atque aliorum medicorum loci obscuri & implicati illustrantur & explicantur: quae quidem omnia, omnibus verae et incorruptae medicinae studiosis tum utilissima, tum necessaria sunt. Eiusdem Annotationes in Antonii Musae Brasavolae Simplicium medicamentorum examen* (Basel: apud Mich. Ising[rinium], [ca. 1550]).

24. Lange frequently referred to "Plutarchus in Symposio" or "Plutarchus in Symposiis," such as in Lange, *Epistolae* (1605), 1.74, 76, pp. 405, 415 (= *Miscellanea* [1554], pp. 332, 341); idem, *Epistolae* (1605), 2.3, 13, 24, 28, 36, 43, 44, 51, pp. 498, 552, 598, 617, 657, 687, 692, 729 (= *Miscellanea* [1560], pp. 25, 69, 105, 123, 152, 177, 189, 211), and elsewhere. On the reception of Plutarch's *Moralia* (including the dialogues that comprise the *Symposium*) during the Renaissance in Italy and in sixteenth-century Germany and, more specifically, the popularity of Plutarch in the circle of Melanchthon, see Rudolf Hirzel, *Plutarch* (Leipzig, 1912), 102–21, especially 112 (my thanks to Anthony Grafton for this reference).

25. Lange, *Epistolae* (1605), 1.64–68, pp. 316–68 (= *Miscellanea* [1554], pp. 261–303). Nutton, "Humanist Surgery," 302n68, also suggests "literary artifice."

26. Lange, *Epistolae* (1605), 1.29, 1.50–52, pp. 117–23, 222–42 (= *Miscellanea* [1554], pp. 97–102).

27. On the ruins of Roman baths: Hasenclever, "Die tagebuchartigen Aufzeichen," 423; Lange, *Epistolae* (1605), 1.50, p. 223 (= *Miscellanea* [1554], p. 184). On his visit to the supposed tomb of Roland: Hubertus Thomas Leodius, *De Palatinorum origine . . . commentatio* (separately paginated, 6–7), in Marquard Freher, *Originum Palatinarum pars prima* [Heidelberg:] typis Gotthardi Voegelini, 1612–13).

28. Lange, Epistolae (1605), 2.3, pp. 497–507 (= Miscellanea [1560], pp. 24–32): "De veterum medicinae, aliarumque facultatum bibliothecis, et chalchographia: ad illustrissimum principem Otthenricum, Com. Palat. Electorem." The sources indicated in Lange's account of ancient libraries include Pliny, Herodotus, Aulus Gellius, Galen, Tertullian, Justin Martyr, Josephus, Eusebius, Plutarch, Suetonius, and Livy. Idem, *Epistolae* (1605), 2.1, pp. 467–71 (= *Miscellanea* [1560], pp. 1–4), is also an encomium to Ottheinrich (1502–1559). On the role of Ottheinrich (who succeeded his uncle Friedrich II and ruled as Elector Palatine 1556–59) in assembling the celebrated Palatine library, see *Bibliotheca Palatina: Katalog zur Ausstellung vom 8. Juli bis 2. November 1986, Heiliggeistkirche Heidelberg, Textband*, ed. Elmar Mittler et al. (Heidelberg: Edition Braus, [1986]), especially 203–4; Karl Schottenloher, *Pfalzgraf Ottheinrich und das Buch: Ein Beitrag zur Geschichte der evangelischen Publizistik. Mit Anhang: Das Reformationsschrifttum in der Palatina*, Reformationsgeschichtliche Studien und Texte, Heft 50/51 (Münster: Aschendorff,

1927). On Renaissance accounts of ancient libraries, see Paul Nelles, "The Renaissance Ancient Library: Tradition and Christian Antiquity," in *Les humanistes et leur bibliothèque/Humanists and Their Libraries*, ed. Rudolf de Smet (Leuven: Peeters, 2002), 1579–73. On Justus Lipsius, *De bibliothecis syntagma* (Antwerp, 1607), and other early accounts of libraries, see also Thomas D. Walker, "Justus Lipsius and the Historiography of Libraries," *Libraries & Culture* 26 (1991): 49–65.

29. Lange, *Epistolae* (1605), 2.20–22, pp. 578–90 (= *Miscellanea* [1560], pp. 89–98); idem, *Epistolae* (1605), 2.21, p. 587, includes a citation of the French botanical author Jean Ruel (1479–1537).

30. Lange, *Epistolae* (1605), 2.20, pp. 583–84 (= *Miscellanea* [1560], pp. 93–94): "Atqui in mense Martio ad Hercyniam sylvam, et iuxta ultrarhenana Palatinorum oppida, Creutzenach et Ingelheim, intra Bingen et Maguntiam, locis incultis, enascitur frutex dodrantalis, ternis ut plurimum coliculis, per ambitum foliis abrotani vestitis; in quorum singulis foliosis capitulis, appetente veris initio, singuli flores instar florum camomillae flavescunt, quibus deciduis, succrescit folliculus, seminibus vulgaris diptami albi plenus, et pariter nigris. Radices porro plures nigras, graciles et complicatas habet, una ex ceposa infimae caulis radice prodeuntes, radicibus veratri nigri per omnia sapore et viribus purgatoriis pares: quas nostri et Italorum quoque agyrtae *rhizotomoi*, id est radiciseci, pharmacopoeis nostris et Venetorum pro veratri nigri radicibus vendunt: nec aliis eruditi Italorum medici, quod ego viderim, utuntur. Quibus et veterinarii nostri mulomedici equos expurgant: ex his quoque clysteria alvi decoquunt. Hanc esse Dioscoridis et Theophrasti nostri elleborinam, quam *to poarion*, id est herbulam appellat, et Plinii epipactida, et radices eius esse veratri nigri antiquorum succidaneas, arbitror."

31. Vivian Nutton, ed., *Medicine at the Courts of Europe, 1500–1837* (London: Routledge, 1990). In a recent collection of studies of knowledge at European courts between 1400 and 1600 in *Micrologus* 16 (2008), *I saperi nelle corti/Knowledge at the Courts*, no fewer than seven of the twenty-five articles relate to medicine: M. Nicoud, "Les savoirs diététiques, entre contraintes médicales et plaisirs aristocratiques," 233–55; R. Martorelli Vico, "Fondamenti biologici di uno *Speculum principis*: Il *De regimine principum* di Egidio Romano," 257–70; D. Jacquart, "Naissance d'une pédiatrie en milieu de cour," 271–94; M. Ferrari, "Ordine da servare nella vita ed emploi du temps: Il ruolo pedagogico del medico in due corti europee tra '400 e '600," 295–312; G. Zuccolin, "Sapere medico e istruzioni etico-politiche: Michele Savonarola alla corte estense," 313–26; F. Bacchelli, "Antonio Musa Brasavola archiatra di Ercole II duca di Ferrara," 327–46; and V. Segre, "La medicina e la cultura della corte nei *Tacuina sanitatis* illustrate," 347–72.

32. Examples include the influence enjoyed by Michele Savonarola as an adviser at the Estense court (see Zuccolin, "Sapere medico e istruzione etico-politiche") and the double role of Wolfgang Lazius as physician and historian at

the Austrian Hapsburg court (see Nancy G. Siraisi, *History, Medicine, and the Traditions of Renaissance Learning* [Ann Arbor: University of Michigan Press, 2007], chapter 6); on the opposite side there is Cardano's narrow escape from an enraged noble father whose son died under Cardano's care (see Nancy G. Siraisi, *The Clock and the Mirror: Girolamo Cardano and Renaissance Medicine* [Princeton: Princeton University Press, 1997], 174).

33. Lange, *Epistolae* (1605), 1.13–14, pp. 55–65 (= *Miscellanea* [1554], pp. 46–54).

34. On the basis of one of Lange's *consilia*—Lange, *Epistolae* (1605), 1.21, pp. 89–93 (= *Miscellanea* [1554], pp. 74–77): "De morbo virgineo epistola"—he has often been credited with a pioneering description of the disease known to the early modern period as green sickness and from the seventeenth to nineteenth centuries as chlorosis. Helen King, *The Disease of Virgins: Green Sickness, Chlorosis and the Problems of Puberty* (London: Routledge, 2004), chapter 2, shows that this is a misapprehension, in that Lange based his exposition on his own interpretation of Hippocrates read in Calvi's early sixteenth-century Latin edition and did not regard the disease as new; King edits the letter at 142–43.

35. Lange, *Epistolae* (1605), 1.18, pp. 76–81 (= *Miscellanea* [1554], pp. 63–68): "An curandae pestis methodus tam per phlebotomiam quam pharmaca evacuationem exigat, et quae sit alteri praemittenda," with reference at the end to his successful treatment of many—"ingens aegrorum numerus"—plague patients ("pestis" may, but does not necessarily, imply bubonic plague). The plague outbreak was possibly that of 1530–31 or 1547; see Rosemarie Jansen and Hans Helmut Jansen, "Die Pest in Heidelberg," in *Semper Apertus: Sechshundert Jahre Ruprecht-Karls-Universität Heidelberg, 1386–1986*, ed. Wilhelm Doerr et al., vol. 1, *Mittelalter und Frühe Neuzeit, 1386–1803* (Berlin: Springer-Verlag 1985), 371–98, at 376 (with no mention of Lange). Lange, *Epistolae* (1605), 2.39 (on stillbirth, or mola, suffered by the wife of the prefect of the village of Kalbe), pp. 667–73 (= *Miscellanea* [1560], pp. 160–64); idem, *Epistolae* (1605), 1.81 (on insect pests in pine trees), pp. 446[misprint for 447]–53 (= *Miscellanea* [1554], 368–72).

36. Lange, *Epistolae* (1605), 1.32 and 1.45 (on illnesses and treatment of Ottheinrich and Friedrich), pp. 130–34, 200–204 (= *Miscellanea* [1554], pp. 107–10, 165–68); idem, *Epistolae* (1605), 1.54 (on the infant son of Patrocles), pp. 250–56 (= *Miscellanea* 1554, pp. 206–10); idem, *Epistolae* (1605), 1.58 (on Kress), pp. 278–81 (= *Miscellanea* 1554, pp. 228–31); also see Jonathan W. Zophy, "Lazarus Spengler, Christoph Kress, and Nuremberg's Reformation Diplomacy," *Sixteenth Century Journal* 5 [1974]: 35–48); idem, *Epistolae* (1605), 2.45 (on Arnoldus), pp. 694–99 (= *Miscellanea* 1560, pp. 182–86); idem, *Epistolae* (1605), 3.3 (on the fall of the drunken youth), pp. 881–86. For examples of discussions of foodstuffs, eating, and drinking, see idem, *Epistolae* (1605), 1.55–62, pp. 256–304 (= *Miscellanea* [1554], pp. 210–50), and others; idem, *Epistolae* (1605), 2.51 (to Christoph Prob), pp. 726–31 (= *Miscellanea* [1560], pp. 209–12).

37. Lange, *Epistolae* (1605), 1.44 (discussion during a hunting trip with the ruler's cook and steward about the harmfulness of lead pipes and cooking vessels), pp. 197–200 (= *Miscellanea* [1554], pp. 163–65). Idem, *Epistolae* (1605), 1.69 (whether, by eating poison, immunity may be acquired), pp. 368–71 (= *Miscellanea* [1554], pp. 303–6), is addressed to Pamphilus, who was present when the hunters of the Count Palatine, whom Lange had taught about antidotes, brought doves killed with henbane, partridges fattened on hellebore, and fish caught with nux vomica, as Pamphilus had warned Lange to be careful about eating them.

38. Lange, *Epistolae* (1605), 3.1, pp. 777–877: "De secretis remediis, ad Georgium Wirth medicum Caesaream consobrinum epistola." This "letter" is actually a pharmacological treatise in twelve chapters, in which recipes are divided according to type (pills, powders, ointments, etc.). For recipes attributed to Lange in medical miscellanies among the Heidelberg Codices Palatini germanici, see index entries under his name in Miller and Zimmerman, *Codices Palatini germanici* (2005), and idem, *Die medizinischen Handschriften*. In addition, Heidelberg Cod. Pal. germ. 193, which contains at 68r–74r a *consilium* for Ludwig VI (1539–1583; reigned 1576–83) attributed to Johann Kraus, Thomas Erastus, and Johann Lange (Miller and Zimmerman, *Codices Palatini germanici* (2005), 51; idem, *Die medizinischen Handschriften*, 131), has been digitized and can be seen at http://digi.ub.uni-heidelberg.de/diglit/cpg193/0147. See also Joachim Telle, "Mitteilungen aus dem 'Zwölfbändigen Buch der Medizin' zu Heidelberg," *Sudhoffs Archiv* 52 (1968): 310–40, at 320–21. A letter from Lange to Friedrich III (1515–1576; Elector Palatine, 1559–76), dated 1564 and containing medical advice, survives in Heidelberg Cod. Pal. germ. 8, 34r–37v; see Karin Zimmerman et al., eds., *Die Codices Palatini germanici in der Universität Heidelberg (Cod. Pal. germ. 1–181)* (Wiesbaden: Dr. Ludwig Richert Verlag, 2003), p. 17, no. 16, http://diglit.ub.uni-heidelberg.de/diglit/zimmermann2003. I have not seen the originals of these Heidelberg manuscripts. Further research into Lange's recipes might reveal more about his medical practice. In particular, a comparison of the manuscript recipes attributed to him in volumes among the Codices Palatini germanici with the Latin collection of secret remedies he prepared for Georg Wirth (Lange, *Epistolae* [1605], pp. 777–877) could be of interest. A very few Latin recipes attributed to Lange are also to be found in medical miscellanies among the Codices Palatini latini; see Ludwig Schuba, ed., *Die medizinischen Handschriften der Codices Palatini Latiniin der Vatikanischen Bibliothek* (Wiesbaden: Dr. Ludwig Reichert Verlag, 1981), 346, 433, 484.

39. Lange, *Epistolae* (1605), 2.23, p. 593 (= *Miscellanea* [1560], pp. 101): "Alii sunt et dicuntur receptarii, qui a pharmacopoeis et chymistis absque ulla ratione integros receptorum fasciculos colligunt et recipiunt, quibus plaustra onerare possent: qui neglecta morborum per causas naturales cognitione et dinotione, sola experimenta consectantur . . . At recta morbi medicamenta, non pharmacopoeus aut

chymista, sed recta morbi, et illius causae per signa pathognomica cognitio, indicat et suppeditat: quam absque theorematum philosophiae naturalis doctrina tibi
comparare aegre poteris."

40. For changes affecting the university, see Joachim Telle and Wilhelm
Kühlmann, "Humanismus und Medizin an der Universität Heidelberg im 16.
Jahrhundert," in Doerr et al., *Semper Apertus*, 1:255–89, at 255–59; Elmar Wadle,
"Ottheinrichs Universitätreform und die Juristische Fakultät," in Doerr et al.,
Semper Apertus, 1:290–313; Christopher J. Burchill, "Die Universität zu Heidelberg und der 'fromme' Kurfürst: Ein Beitrag zur Hochschulgeschichte im wer-
denden konfessionellen Zeitalter," in Doerr et al., *Semper Apertus*, 1:231–54. On
the stages of the Reformation in the Palatinate, see Euan Cameron, *The European Reformation* (Oxford: Oxford University Press, 1991), 268–69, 370–71. For the
emergence and political consequences of the Palatinate as a center of militant Calvinism, see Claus-Peter Clasen, *The Palatinate in European History* (Oxford: Basil
Blackwell, 1963).

41. Telle and Kühlmann, "Humanismus und Medizin," 255–59; Nutton, "Humanist Surgery," 95; Eberhard Stübler, *Geschichte der medizinischen Fakultät der
Universität Heidelberg, 1386–1925* (Heidelberg, 1926), 33–42. For Lange's reform
proposal of 1545, see August Thorbecke, ed., *Statuten und Reformationen der Universität Heidelberg vom 16. bis 18. Jahrhundert* (Leipzig, 1891), 355–58; for Ottheinrich's statutes, see the medical faculty section, Thorbecke, *Statuten*, 76–91. The
date of composition of Lange's *Symposium* is established by the colophon, 110[mis-
print for 100]: "anno a Deiparae partu 1547, die mensis Martii 14, finitum."

42. Compare Lange's 1545 proposal (in Thorbecke, *Statuten*, 356) and the 1558
statutes (in Thorbecke, *Statuten*, 78–79). On Erastus's role in the 1558 statutes for
the faculty of medicine, see Thorbecke, *Statuten*, 78n2–3, and Stübler, *Geschichte
der medizinischen Fakultät*, 33; for the influence of Erastus as professor of medicine,
anti-Paracelsian, and religious controversialist at the University of Heidelberg, see
Telle and Kühlmann, "Humanismus und Medizin," 265–71. The 1558 statutes apparently allowed some provision for use of sections of Avicenna's *Canon* and Rasis's
Almansor; see Thorbecke, *Statuten*, 78–79n3. On Lange's interest in the reform of
surgical education, see especially Nutton, "Humanist Surgery," 91–96.

43. Thorbecke, *Statuten*, 358; ibid., no. 99, at 89–90. Contrast the situation at
the University of Padua, where, for most of the sixteenth century, some Jewish medical practitioners were able to study—twenty-nine are recorded as having done so
between 1520 and 1605 (David B. Ruderman, *Jewish Thought and Scientific Discovery
in Early Modern Europe* [Detroit: Wayne State University Press, 2001], 105).

44. Lange, *Epistolae* (1605), 2.18, pp. 567–73 (= *Miscellanea* [1560], pp. 80–85),
at 567: "Memorabilis sane . . . fuit olim imperante Antonino Caesare, vetus Romane
urbis medicorum consuetudo, qui dum sol ad occasum vergeret, apud templum
Pacis, unde bibliothecae Romanorum non longe aberant, sedulo conveniebant,

quisque de suae sectae thematibus disputaturus. Horum exemplo Heidelbergensis Academiae philiatri admoniti, non raro a coena apud Chilianum pharmacopoeum e regione templi Sancti Spiritus conveniunt." Rascalon, described as an "eruditionis adolescens," is one of the participants in the dialogue in idem, *Epistolae* (1605), 2.18; in addition, ibid., 2.39 and 2.49, pp. 667–73, 715–23 (= *Miscellanea* [1560], pp. 160–64, 200–206), are addressed to him (termed, in 2.49, "candidatorum medicinae decus"). On Rascalon, see Didier Kahn, *Alchimie et Paracelsisme en France à la fin de la Renaissance (1567–1625)* (Geneva: Librairie Droz, 2007), 96, 107; and for Rascalon's epitaph for the learned Protestant refugee Olympia Fulvia Morata, see Adam, *Vitae germanorum medicorum*, 82. Lange, *Epistolae* (1605), 2.60 (to Lotichius, 1558 ["annus iam . . . praeteriit, quo Heidelbergam accersitus est"]), pp. 764–75 (= *Miscellanea* [1560], pp. 239–46), denounces medical astrology. On Lotichius (1528–1560), who came to Heidelberg as professor of medicine and botany in 1557, see Adalbert Elschenbroich, "Lotichius, Petrus Secundus," *NDB* 15 (1987): 238–41, www.deutsche-biographie.de/sfz54412.html, and Telle and Kühlmann, "Humanismus und Medizin," 259–65.

45. Notably Ludwig V (1478–1544; Count Palatine of the Rhine, 1508–44), who collected and transcribed a very large number of medical recipes; Friedrich II, who ordered his predecessor's collection to be bound (see Joachim Telle, "Mitteilungen aus dem 'Zwölfbändigen Buch der Medizin' zu Heidelberg," *Sudhoffs Archiv* 52 (1968), 310–40); and Ludwig VI.

46. Lange, *Epistolae* (1605), 2.12, pp. 548–50 (= *Miscellanea* [1560], pp. 65–66): "Quum haec, Alceae graviter aegrotante principe meditaremur, nobiscum generosis stemmate Comitibus una discumbentibus, ut temporis fastidium falleremus, admirandas muliebris picae . . . historias convivae recensebant . . . Lembergi vero, in amoeneo Schlesiae et regali oppido, in quo ego natus sum, sacerdotem lotis et nudis pedibus egressum balneo, inque plantis de more crepidas deferentem, mulier gravida subsequuta, pedum desiderio capta, procidit a tergo in genua: ac ambabus manibus crure alter apprehenso, de calce sacerdotis mordicus dentibus frustum evulsit, nihil illius clamorem expavescans, quo deorum hominumque fidem et subsidium is implorabat." The anecdote reflects Lange's Lutheran adherence, in that it was presumably intended to bring ridicule on the Catholic cleric as well as the woman.

47. Lange, *Epistolae* (1605), 2.27, pp. 604–15 (= *Miscellanea* 1560, pp. 110–10 [misprint for 119]). At the Diet of Speyer, "a partu Christiparae virginis anno supra mille quingentos quadragesimo secundo" (p. 604), Lange discussed the case of the fasting girl with a colleague from Leipzig (on this case regarding Margarethe Weiss, see Anne Jacobson Schutte, *Aspiring Saints* [Baltimore: Johns Hopkins University Press, 2001], 139–41). Lange's medical attendance on Friedrich at the Diet of Speyer, recorded in Lange, *Epistolae* (1605), 1.45, pp. 200–4 (= *Miscellanea* [1554], pp. 165–68), where a consultation among attending physicians about Friedrich's illness and

treatment was held in a church, presumably took place during the Diet of 1544, as Friedrich is described as Count Palatine and Elector; idem, *Epistolae* (1605), 1.55–56, pp. 256–73 (= *Miscellanea* [1554], pp. 210–24), present dialogues on the topics "should lunch be lighter than supper" and on different kinds of bread and grain among the attending physicians (some from Thuringia and Austria) of the dukes and emperor gathered for the Diet of Augsburg (1548); idem, *Epistolae* (1605), 1.58, pp. 278–81 (= *Miscellanea* [1554], pp. 228–31), recalls conversations at Nuremberg when Lange was called there to treat Christoph Kress.

48. Lange, *Epistolae* (1605), 1.64–67, 71, pp. 316–56, 378–92.

49. On these members of the Wirth family, see Adam, *Vitae germanorum medicorum*, 13–14, and Reusner, "Praefatio," in Lange, *Epistolae* (1589). Lange, *Epistolae* (1605), 1.79, pp. 429–36 (= *Miscellanea* [1554], pp. 353–58): "De naturali vitae periodo, et an prolongari possit epistola . . . Ad Petrum Wirth theologum consobrinum suum Mecoenatem." For ancient sources of and commentators on the device for abbreviating long chronologies and Renaissance uses of it, see Anthony Grafton, "Tradition and Technique in Historical Chronology," in *Ancient History and the Antiquarian: Essays in Memory of Arnaldo Momigliano*, ed. M. H. Crawford and C. R. Ligota (London: Warburg Institute, University of London, 1995), 15–31, at 26–28.

50. Lange, *Epistolae* (1605), 1.5, pp. 24–29; idem, 2.27–28, pp. 609–21. On Heinrich Stromer of Auerbach, see chapter 1. For Johann Reusch von Eschenbach, another humanist and physician at the University of Leipzig in the second decade of the sixteenth century, see Leonhardt, "Classics as Textbooks," 99. Lange, *Epistolae* (1605), 1.35, p. 150 (= *Miscellanea* [1554], pp. 124–33), is addressed to an unnamed colleague or former student from Leipzig.

51. Lange, *Epistolae* (1605), 2.39, pp. 667–73 (= *Miscellanea* [1560], 160–64), is addressed to Gasser (1505–1577), who settled in Augsburg in 1546; for his career and writings, see Friedrich Blendinger, "Gasser (Gassarus), Achilles Pirminius," *NDB* 6 (1964): 79–80, www.deutsche-biographie.de/sfz19959.html. Lange, *Epistolae* (1605), 2.20–22, pp. 578–90 (= *Miscellanea* [1560], pp. 89–98), are addressed to Georg Forster (1510–1568); on Forster, who was close to Luther's circle and also spent part of his career at Heidelberg before settling at Nuremberg, see Gerald Gillespie, "Notes on the Evolution of German Renaissance Lyricism," *MLN* [*Modern Language Notes*] 81 (1966): 437–62, at 441; on Johannes Moibanus (1527–1562), son of the Lutheran pastor Ambrosius Moibanus and practitioner of poetry, music, and painting as well as medicine, see Adam, *Vitae Germanorum medicorum*, 120–27. According to Adam, Johannes Moibanus was a friend of both Crato and Gasser.

52. Lange, *Epistolae* (1605), 2.53, pp. 741–44 (= *Miscellanea* [1560], pp. 221–23). But Lange's connection with Bech (1521–1560) may originally have been a Leipzig one, since Lange explained that he called Bech *perdilecte* both because of his learn-

ing and because he had been the roommate of Georg Wirth, Lange's cousin, at the University of Leipzig. See also Reusner, "Praefatio," in Lange, *Epistolae* (1589), α4v, and Adam, *Vitae germanorum medicorum*, 13–14.

53. Lange, *Epistolae* (1605), 2.29, pp. 621–24 (= *Miscellanea* [1560], pp. 124–36).

54. Lange, *Oratio Ioannis Langij Lembergij.*

55. Hasenclever, "Die tagebuchartigen Aufzeichen," 399.

56. Lange, *Epistolae* (1605), 1.71, p. 379 (= *Miscellanea* [1554], p. 312).

57. See Cameron, *European Reformation*, 268–89, 370–71.

58. The colophon establishes the date of composition (1547) of Lange's *Symposium* (published in 1554). The dedicatee is Eberhard, count of Erbach, archipincerna (arch-cupbearer—a hereditary office of the counts of Erbach) to the Elector Palatine of the Rhine.

59. Lange, *Symposium* (1554), at p. 88: "Tali aqua non modo animarum labes, sed quamvis morbidam corporis luem expiari posse populo persuadetis . . . Huius aspergine corpora et sepulchra defunctorum ad vesperam lustrantur . . . Hos impios ceremoniarum ritus vos ex veterum magia et superstitione desumpsisse"; and at p. 92: "Tota vestra religio idolatriae cultu defoedata, medicinam redolet, ob pecuniae aucupium. Ingredere, inquam, quaeso hoc Augustum Vindelicorum templum, ut videas quanta donaria ex cera, ferro, argento et auro conflate divorum simulacris, quae adorare nefas, ab aegris dicata sunt."

60. Lange, *Epistolae* (1605), 1.33, pp. 134–43 (= *Miscellanea* [1554], pp. 111–18): "De magia et rerum naturalium sympathia"; Homer, *Odyssey* 11; 1 Samuel 28. The bibliography of Renaissance occultism and natural magic is very extensive; for a concise account, see Brian P. Copenhaver, "Astrology and Magic," in *The Cambridge History of Renaissance Philosophy*, ed. Charles B. Schmitt et al. (Cambridge: Cambridge University Press, 1988), 264–300. The idea that magic had originally been a pure and benign form of knowledge that subsequently became corrupted was also espoused by some medieval Jewish authors; see Ruderman, *Jewish Thought and Scientific Discovery in Early Modern Europe*, 33–41. For the idea that what appeared to be ghosts of the dead were really demonic imitations, see Stuart Clark, *Vanities of the Eye* (Oxford: Oxford University Press, 2007), 236–65.

61. Lange, *Epistolae* (1605), 1.34, pp. 143–50 (= *Miscellanea* [1554], pp. 118–24), continues the same general theme as idem, *Epistolae* (1605), 1.33, pp. 134–43 (= *Miscellanea* [1554], pp. 111–18), discussing ligatures, amulets, and characters and distinguishing between acceptable and unacceptable varieties: "Proinde de his periaptis et amuletis quae ratione nituntur naturali fidem non derogamus . . . Quare merito aniles illos versiculos, ex sacrae scripturae verbis, aut barbaris daemonum nominibus concinnatos, una cum divo Augustino despuimus" (pp. 144–45). On Solomon's and Moses's knowledge of magic, Lange cited Josephus, *Antiquitates Judaicae* 8.2; on this and other testimonies to Solomon as a magician, see Sarah Iles Johnston, "The *Testament of Solomon* from Late Antiquity to the Renaissance," in *The Metamorphosis of Magic from Late Antiquity to the Early Modern Period*, ed.

Jan N. Bremmer and Jan R. Veenstra (Leuven: Peeters, 2002), 35–49; for the Jewish tradition that Solomon composed a (probably magical) book of remedies, see Ruderman, *Jewish Thought and Scientific Discovery in Early Modern Europe*, 379, and David J. Halperin, "*The Book of Remedies*, the Canonization of the Solomonic Writings, and the Riddle of Pseudo-Eusebius," *Jewish Quarterly Review* 72 (1982): 269–92. Lange returned to the subject of sympathies and antipathies in nature and the benign theurgic magic of Moses, Solomon, and other ancient magi (including the Magi of Epiphany) in several other letters: *Epistolae* (1605), 2.10, pp. 539–44 (=*Miscellanea* [1560], pp. 58–61), at p. 542: "Moysem, vatem naturalis et verae magiae peritissimum"; idem, *Epistolae* (1605), 2.18, pp. 567–73 (=*Miscellanea* [1560], pp. 80–85), at pp. 569–70: "Quod non alia accidit ratione, quam (ut Plato in Timaeo ait) quod cuilibet rerum speciei peculiate astrum, ut polus magneti, et cuilibet astro peculiaris intelligentiae daemon, ut homini cuilibet suus genius, dicatus sit. Quicunque horum sympathiae consensum, et antipathiae dissidium noverit, ille admiranda per herbas et lapides, horumque occultas vires et spiritus producet magiae opera"; and especially idem, *Epistolae* (1605), 2.46, "De magica Aegptiorum medicina, ac magiae speciebus," pp. 700–706 (=*Miscellanea* [1560], pp. 187–92).

62. See introduction to this volume, and Mondella, *Epistolae medicinales* (ca. 1550), epistola 26, pp. 364–75.

63. Lange, *Epistolae* (1605), 2.49, pp. 715–23 (=*Miscellanea* [1560], pp. 200–206), at pp. 717–18: "Quis enim rationis compos non irrideat velut anilia deliramenta, in achate coniugium concordiam et castitatem, in topazio sinistri brachii armillis inserto securitatem: et Neptuni imagine hycintho insculpta aequoris tranquillitatem: et magnetem pulvinari subditam, uxoris in somniis prodere adulterium: alias sexcentas huius generis nugas? Quas ne assis quidem facio. Aliae vero sunt in arte curandi morbos gemmarum facultates, experientia repertae, et usu peritorum approbatae . . . Atqui quum horum causas effectuum reddere nequimus, oportet nos rationis inopes, agyrtarum more, ad caecas formae specivocae proprietates, tanquam ad sacram ancoram et imperitorum asylum, vel ad certo imperscrutabiles siderum operationes et effectus, una cum Hermete et philosophis Indorum et Aegyptiorum confugere . . . Aut si mavis, cum Galeno nostro ad rerum simplicium . . . vaporum effluvium confugiamus . . . Sic Galenus expertus est, smaragdum, quem lapidem viridum appellat, supra orificium ventriculi pro amuleto appensum, illud corroborare [in margin: Galen. lib. 9 simpli. phar.]."

64. Lange, *Epistolae* (1605), 1.38, "De prodigiis et daemonum in morbis praestigiis," pp. 166–69 (=*Miscellanea* [1554], pp. 138–40), at pp. 166–67: "Plurima sane, mi Petre, admiratione digna quotidie in morbis et animalium generatione, quorum causas vix ratione ulla assequi possumus, praestigiis similima, vel daemonum astu etiam, non raro naturae viribus accidere videmus: adeo etiam natura Synesio et Plotino plerumque videtur esse maga. Nam ludit, et crebro, in humanis divina sapientia rebus . . . Nihil igitur mirum, et steatomatum, et atheromatum,

aliorumque apostematum concavitatibus carne exesa, de viscosis et putridis humoribus ac fibris calore putredinali siccatis res lapidibus, tophis, harenis, urceolorum testis, lignis, carbonibus, capillis, et amurcae similes nasci, quae vulgus incantationi et praestigiis daemonum accepta refert: cum simillima astu daemonum in corporibus aegrorum praecipue maligno obsessis spiritu, plerumque accidere, etiam nostra aetate viderimus." Lange returned to the subject of demons and demonic possession in idem, *Epistolae* (1605), 2.34, pp. 643–50 (= *Miscellanea* [1560], pp. 342[misprint for 142]–147): "An daemones humana corpora morbis affligant, exorcismo obtemperent: qua denique methodo tales morbi curentur," responding to a report of cases of possession and exorcisms in Rome (his answer is yes, and exorcism is recommended for cases that do not respond to other treatment), and idem, *Epistolae* (1605), 2.35, pp. 650–56 (= *Miscellanea* [1560], pp. 147–51): "An animae, vel daemones obsessorum, ex auctorati procellas in aere et undis continent," with a reference to spirits in mines, accompanied by a citation of Agricola at p. 653 (the question is again answered positively by Lange, though with somewhat less certainty). But at idem, *Epistolae* (1605), 2.34, p. 647 (= *Miscellanea* [1560], pp. 145): "Pomponatius, meus Bononiae in philosophia praeceptor, tradit, priscos exorcistas (quos praecantatores appellant) ante coniurationem, obsessorum corpora ab atra bile expurgasse." In Pietro Pomponazzi's *De incantationibus* (1520), as is well known, Pomponazzi insisted that phenomena popularly attributed to demons were the result of natural (including occult or astral) causes.

65. Nutton, "Humanist Surgery."

66. Lange, *Epistolae* (1605), 1.2, p. 13 (= *Miscellanea* [1554], pp. 10–11): "Deinde quum magistratus et proceres, accepta . . . parva pecunia, libere in corpora subditorum grassari, eosque non tondere, sed deglubere permittant: hac magistratum indulgentia effectum est, ut nulla sit lex (teste Plinio) quae capitalem medicorum inscitiam puniat: solique medico hominem occidere et Iudaeis in corporibus christianorum experimenta per mortes agere, impune liceat. Hac impunitate freti monachi, apostatae, agyrtae, ambubaiarum collegia, balatrones, anus incantatrices, denique carnifices . . . in hanc sacram medicinae professionem gregatim irrepserunt: quam suis incantationibus, superstitionum praestigiis et nugis, adeo defoedarunt, ut nulla sit artium, quae ex illorum ignorantia tot sit superstitionibus et erroribus referta." There are many similar examples. For Lange's proposal for university reform and the section of it adopted at Heidelberg, see Thorbecke, *Statuten*, 355–60 and 89–90; and for his attack on Jewish practitioners, see Lange, *Symposium* (1554), pp. [2], 63, 82.

67. Lange, *Symposium* (1554), p. [2]: "Choriambicum Asclepiadeum in Iudaeos medicos authore Iohanne Langio"; ibid., p. 63: "dic mihi, per tuum te obsecro, Aesculapium qui tum sit, ut . . . Iudaei . . . apud principes et episcopos, adde et proceres, nobilitatis stemmate claros, cum tanta ostentatione et foenore medicinam profiteantur?" Ibid., p. 82, similarly; *agyrtae* (wandering empirics) are also

condemned (e.g., ibid., p. 65), but with less ferocity. For the concluding recommendations, see ibid., p. 98. For the attacks on Catholic religion (in addition to, and as distinct from, medical practice by Catholic clergy) in this treatise, see ibid., p. 88.

68. Lange, *Epistolae* (1605), 1.71, pp. 378–92 (= *Miscellanea* [1554], pp. 311–22): "An Iudaei sanguine humano utantur." The subsequent letter in idem, *Epistolae* (1605), 1.72, pp. 393–400 (= *Miscellanea* [1554], pp. 323–29): "Veteres ex Aegypto philosophos divina herbarum medicamenta, magica et anili superstitione infamasse turpiter," continues the theme of the history of magic and the responsibility of the Jews in disseminating, and corrupting, magic they had learned in Egypt.

69. Lange, *Epistolae* (1605), 1. 71, p. 378: "Tridentum, Simonis pueri martyrio ac Arno flumine celebre, devenimus . . . At enim, [hospes] ait, Langius vetus meus etiam olim conviva et hospes, Mantuae, Bononiae, Venetiis, Iudaeis, ut illorum vel in ceremoniis, aut medicamentis, si qua haberent, arcana, exploraret, diu familiariter conversatus est. is ergo veram vos causam edocere potuit." On the widespread dissemination of material about the Trent ritual murder trials in the German lands and its effect, see R. Po-Chia Hsia, *The Myth of Ritual Murder: Jews and Magic in Reformation Germany* (New Haven: Yale University Press, 1988), 42–50; and on the trial itself, idem, *Trent 1475: Stories of a Ritual Murder Trial* (New Haven: Yale University Press, 1992).

70. Lange, *Epistolae* (1605), 1.71, pp. 379–80: "Viginti enim praeterierunt anni, quum ego adolescens Bononiam per Mirandulam proficiscerer. Aderat ibi tum comitate morum et omnifariae eruditionis splendore clarissimus princeps, Ioannes Franciscus Picus, Mirandulae comes et Concordiae dominus illustris: qui quum ab hospite accepisset, nos ex Germania advenas: mox a nobis per nunciam accersitis, primum omnium de vita et conditione Ioannis Capnionis sciscitabatur: et an controversia eius cum monachis studiorum osoribus, per iudicem aut Caesarem transacta esset. Cui respondi: intempestivam Capnionis mortem contentionem illam diremisse. Tum ingemiscens Mirandulae Comes: Ah quid audio? Interiit certe, inquit, triplicis linguae culmen, et Germaniae decus: cuius opera factum, ut Tyberis defluxerit in Rhenum: et sancta illa ac vetustate venerabilis lingua Hebraica, olim vaga et incondita, nunc auctore Capnione certis constricta legibus, certa dicendi scribendique norma facile omnibus facta est obvia. Mehercle obitu huius viri Hebraismus multum detrimenti accepit. Tandem ego. Probe, O Comes, hunc virum depinxisti, inquam. Siquidem omnes docti et boni viri, praeter monachos, quibus cum literis vel Musis nihil commune est, huius mortem deplorant . . . Sed O Comes, . . . libenter a tua eruditionis omnifariae excellentia percunctaremur, quae nam sint praecipua Iudaeorum de re medica et secretiori philosophia volumina."

71. Lange, *Epistolae* (1605), pp. 380–88. At 380–81: "Esdras in LXX volumina digessit, quam Cabalam dixerunt: quae ineffabilem de divinitate, et angelicis in-

telligentiis sapientiam, et de rebus naturalibus exactam traditionem continebant. Quam postea iuniores, praeter artem revelationis literarum, mille magicae artis, quam in Egypto didicerant, deliramentis et goetiae vanitatibus conspurcarunt."

72. For Giovanni Pico's account of the cabala, see *Oratio*, in Giovanni Pico della Mirandola, *Commentationes* (Bologna, 1496), 138v–139r, consulted at www.brown .edu/Departments/Italian_Studies/pico/incunab/testo/editio.html; on Christian cabala more generally, see François Secret, *Les kabbalistes chrétiens de la Renaissance* (Paris: Dunod, 1964). For Pico's view on magic, see his *Oratio*, 137v: "Proposuimus et Magica theoremata, in quibus duplicem esse magiam significavimus: quarum altera daemonum tota opere et auctoritate constat, res medius fidius execranda et portentosa. Altera nihil est aliud, cum bene exploratur, quam naturalis philosophiae absoluta consummatio," and his *Apologia*, in Giovanni Pico della Mirandola, *Omnia opera: Con un premessa di Eugenio Garin* (Turin: Bottega d'Erasmo, 1971), vol. 1 (containing a facsimile of part of Pico's *Omnia opera* [Basel, 1572]), 120–24; similarly, see Marsilio Ficino, "Apologia," in his *Three Books on Life*, ed. and trans. Carol V. Kaske and John R. Clark (Binghamton, NY: Center for Medieval and Early Renaissance Studies, 1989), 398; and, with emphasis on the decline and corruption of magic over time, Ioannes Trithemius, letter to Heinrich Cornelius Agrippa, in Heinrich Cornelius Agrippa, *De occulta philosophia libri tres*, ed. Vittoria Perrone Compagni (Leiden: Brill, 1992), 68–71.

73. Lange, *Epistolae* (1605), 1.71, p. 387: "At, ut audio, non desunt qui Iudaeorum impietati patrocinentur, et haec [ritual murder accusations] de Iudaeis conficta fuisse crimina affirmant, ut eorum bona confiscarentur, et iudices ditescerent." Could this be an allusion to Osiander's treatise (written in 1529, published anonymously in 1540) refuting the blood libel (see Hsia, *Myth of Ritual Murder*, 136–43, and Heiko Oberman, *The Impact of the Reformation* [Grand Rapids, MI: Wm. B. Eerdmans, 1994], 99–100)? If so, it suggests that Lange's reworking of memories of an encounter with Pico in the early 1520s was both anachronistic and extensive. Lange, *Epistolae* (1605), 1.71, p. 388: "Tum Geraldus. Haec quae nobis de Iudaeorum magica impietate et eorum goetiae praestigiis, quae sanguinem, daemoniorum philtrum requirunt disseruisti, non multum diversa a Capnione nostro, ex libro de verbo mirifico accepimus." On the actual content of Reuchlin's treatise, see Charles Zika, "Reuchlin's *De verbo mirifico* and the Magic Debate of the Late Fifteenth Century," *Journal of the Warburg and Courtauld Institutes* 65 (1976): 104–38.

74. Richard H. Popkin, *The History of Skepticism from Savonarola to Bayle*, 3rd ed. (Oxford: Oxford University Press, 2003), 20–27; Charles B. Schmitt, *Gianfrancesco Pico della Mirandola (1469–1533) and His Critique of Aristotle* (The Hague: Martinus Nijhoff, 1967).

75. For an analysis of Gianfrancesco Pico's attitude toward witchcraft and of his dialogue *Strix*, in which a skeptic is convinced of witchcraft's reality, see Walter Stephens, *Demon Lovers: Witchcraft, Sex, and the Crisis of Belief* (Chicago: Uni-

versity of Chicago Press, 2002), especially 87–102, and idem, "Gianfrancesco Pico e la paura dell'immaginazione: Dallo scettismo alla stregoneria," *Rinascimento* 43 (2004): 49–74; for Gianfrancesco Pico's denunciations of magic, see D. P. Walker, *Spiritual and Demonic Magic from Ficino to Campanella* (University Park: Pennsylvania State University Press, 2003; originally published London: Warburg Institute, 1958), 146–51. On the younger Pico's attitude toward his uncle, see Brian Copenhaver, "Studied as an Oration: Readers of Pico's Letters, Ancient and Modern," in *Laus Platonici philosophi: Marsilio Ficino and His Influence*, ed. Stephen Clucas, Peter J. Forshaw, and Valery Rees (Leiden: Brill, 2011), 149–98.

76. Gianfrancesco Pico, *De rerum praenotione libri novem. Pro veritate religionis contra superstitiosas vanitates editi . . .* (Strasbourg, 1507), bk. 7, chapter 8, "Adversus nonnullos Haebreorum magos," [Rv v–Rvi r]. Regarding the book of remedies attributed to Solomon, see Ruderman, *Jewish Thought and Scientific Discovery in Early Modern Europe*, 379, and Halperin, "*The Book of Remedies.*"

77. Lange, *Epistolae* (1605), 1.71, pp. 388–92. I owe the suggestion about Paracelsus to David Ruderman; see also Ruderman, *Jewish Thought and Scientific Discovery in Early Modern Europe*, 245, and Harry Friedenwald, "Apologetic Works of Jewish Physicians," in *The Jews and Medicine: Essays*, 2 vols. (Jersey City, NJ: Ktav, 1967; first published Baltimore: Johns Hopkins University Press, 1944), 1:55–56, both quoting from Paracelsus's preface in the Nuremberg edition (1553) of his *Labyrinthus medicorum errantium* (a work written in 1538).

78. For the Christian perception of Jews as engaged in magic, see Hsia, *Myth of Ritual Murder*, passim; regarding Jewish conceptions of magic at that time, see Moshe Idel, "Jewish Magic from the Renaissance Period to Early Hasidism," in *Religion, Science, and Magic in Concert and in Conflict*, ed. Jacob Neusner, Ernest S. Frerichs, and Paul Virgil McCracken Flesher (Oxford: Oxford University Press, 1989), 82–117.

79. "Picus in Epistolis" is cited in Lange, *Epistolae* (1605), 1.71, p. 379; Reuchlin, *De verbo mirifico*, is cited in Lange, *Epistolae* (1605), 1.71, p. 388. Crinitus, cited on the cabala and on magic in ibid., 1.71, pp. 380, 382, is presumably Petrus Crinitus (Pietro Riccio, 1476–1505), who summarized and quoted from the elder Pico's views on cabala in his *Commentarii de honesta disciplina* (first published 1504), bk. 25, chapter 3. For an insightful characterization of Christian Hebraism in the lands of the Holy Roman Empire, see Erika Rummel, "Humanists, Jews, and Judaism," in *Jews, Judaism, and the Reformation in Sixteenth-Century Germany*, ed. Dean Philip Bell and Stephen G. Burnett (Leiden: Brill, 2006), 3–31. I owe the last two references to Anthony Grafton.

80. See Elisheva Carlebach, *Divided Souls: Converts from Judaism in Germany, 1500–1750* (New Haven: Yale University Press, 2001), 170–82. In addition to Margaritha, another author of a similar work contemporary with Lange was Victor von Carben (1442–1515). See also Yaacov Deutsch, "Polemical Ethnographies:

Descriptions of Yom Kippur in the Writings of Christian Hebraists and Jewish Converts to Christianity in Early Modern Europe," in *Hebraica Veritas? Christian Hebraists and the Study of Judaism in Early Modern Europe*, ed. Allison P. Coudert and Jeffrey S. Shoulson (Philadelphia: University of Pennsylvania Press, 2004), 202–33; and Maria Diemling, "Anthonius Margaritha on the 'Whole Jewish Faith': A Sixteenth-Century Convert from Judaism and His Depiction of the Jewish Religion," in Bell and Burnett, *Jews, Judaism, and the Reformation in Sixteenth-Century Germany*, 303–33.

81. In Lange, *Epistolae* (1605), 1.72, pp. 381, 387, 394, references to Schemhamphoras (defined by Lange as nefarious Jewish magic about seventy-two angels or names of God) suggest that he had read one of Luther's anti-Jewish polemics; for Margaritha's influence on this work, see Carlebach, *Divided Souls*, 182.

82. John M. Efron, *Medicine and the German Jews* (New Haven: Yale University Press, 2001), 45, attributes the origins of a specifically German "medical anti-Semitism" to the early modern period. However, as David B. Ruderman correctly points out in his review of this work, hostility toward Jewish medical practitioners occurred much earlier in the Iberian Peninsula (*Jewish Quarterly Review* 92 [2002]: 638–43, at 639). For examples of both ecclesiastical and secular legislation from various parts of southern Europe (dating from the thirteenth and fourteenth centuries) enjoining Christians against consulting Jewish medical practitioners, see Joseph Shatzmiller, *Jews, Medicine, and Medieval Society* (Berkeley: University of California Press, 1994), 90–93; for an example of early fourteenth-century Christian polemic on the subject, see Joseph Ziegler, *Medicine and Religion, c. 1300: The Case of Arnau de Vilanova* (Oxford: Clarendon Press, 1998), 256–58. Friedenwald, "Apologetic Works of Jewish Physicians," 1:31–68, at 31–32, lists a series of sixteenth-century papal prohibitions on the consultation of Jewish physicians by Christians; ibid., at 53–56, for contemporary accusations in the German lands against Jewish physicians.

83. See Rotraud Ries, "German Territorial Princes and the Jews," in *In and Out of the Ghetto: Jewish-Gentile Relations in Late Medieval and Early Modern Germany*, ed. R. Po-Chia Hsia and Hartmut Lehmann (Washington, DC: German Historical Institute, 1995), 215–45.

84. Leopold Lowenstein, *Beitrage zur Geschichte der Juden in Deutschland*, vol. 1, *Geschichte der Juden in der Kurpfalz* (Frankfurt, 1895), 43–48, edits a tax register of Jews living in the Palatinate in 1550 and counts a Jewish population of 155; for discussion, see Dean Philip Bell, "Jewish Settlement, Politics, and the Reformation," in Bell and Burnett, *Jews, Judaism, and the Reformation in Sixteenth-Century Germany*, 421–50, with the figures quoted in my text, which are based on Lowenstein, at 428, 430; and Michael Toch, "Aspects of Stratification of Early Modern German Jewry: Population History and Village Jews," in Hsia and Lehmann, *In and Out of the Ghetto*, 77–89. Bell, "Jewish Settlement, Politics, and the Reforma-

tion," 428, indicates that in 1600 Jews composed about 2 percent of the population of the Empire.

85. The names of about a dozen Jewish practitioners are indexed in the "Register der Rezeptzuträger, Probanden und Gewährsleute," in Miller and Zimmerman, *Codices Palatini germanici* (2005); among them are Alexander and the "Jew from Kreuznach," who contributed recipes—the latter, very numerous recipes—to almost all of the many volumes of Ludwig V's collection. On various regulations, licenses, and taxes of Ludwig V and Friedrich II pertaining to individual Jewishmedical practitioners, see Lowenstein, *Beitrage zur Geschichte der Juden in der Kurpfalz*, 30–35. For relations of Jewish medical practitioners and Christian patients in early modern German lands not restricted to the Palatinate, see Robert Jutte, "Contacts at the Bedside: Jewish Physicians and Their Christian Patients," in Hsia and Lehmann, *In and Out of the Ghetto*, 137–50; and for the summons by Luther's protector, Philip of Hesse, of a Jewish practitioner to treat his invalid sister, see Alisha Rankin, "Duchess Heal Thyself: Elisabeth of Rochlitz and the Patient's Perspective in Early Modern Germany," *Bulletin of the History of Medicine* 82 (2008): 109–44 at 122.

86. For much of the period when Ottheinrich was ruler of Pfalz-Neuburg (before his succession as Elector Palatine in 1556), he was relatively open to Jewish settlement and depended heavily for revenue on the taxation of Jews in his domains (his various cultural projects were very expensive); see Wilhelm Volkert, "Die Juden im Fürstentum Pfalz-Neuburg," *Zeitschrift für Bayerische Landesgeschichte* 26 (1963): 560–605, at 564–7. In 1552, Ottheinrich's attitude changed and he began to expel Jews from Pfalz-Neuburg; his hostility increased in 1556 when he learned that Jews in Heidelberg hoped to prevent his succeeding Friedrich II as Elector Palatine; see ibid., 577–79, and Salo W. Baron, *A Social and Religious History of the Jews: Late Middle Ages and Era of European Expansion, 1200–1650*, vol. 13, *Inquisition, Renaissance, and Reformation* (New York: Columbia University Press, 1969), 256–58.

Chapter 3 · *The Medical Networks of Orazio Augenio*

1. On the distribution of graduate physicians in urban practice in a region of Italy, see Carlo Cipolla, *Public Health and the Medical Profession in the Renaissance* (Cambridge: Cambridge University Press, 1976), "The Medical Profession in Galileo's Tuscany," 67–124, referring primarily to the early seventeenth century.

2. Biographical accounts of Augenio are found in Giovanni Panelli d'Acquaviva, *Memorie degli uomini illustri e chiari in medicina del Piceno, o sia della Marca d'Ancona, 2* (Ascoli, 1757–1758): 191–96 (my thanks to Elisa Andretta for drawing my attention to this work); Loris Premuda, "Augenio, Orazio," *DBI*, www.treccani .it/enciclopedia/orazio-augenio_(Dizionario-Biografico)/; and Alessandro Simili,

"Orazio Augenio da Monte Santo (vita et opere)," *Minerva Medica* 51 (1960): 1613–36.

3. For Augenio's teaching at Macerata, see Antonio Riccoboni, *De gymnasio Patavino . . . commentariorum libri sex* (Padua: apud Franciscum Bolzetam, 1598), 69v. For Augenio's memories of Camerino, see Orazio Augenio, *Epistolarum et consultationum medicinalium prioris tomi libri XII [– alterius tomi libri XII . . .] Hac ed. 5 ab eodem authore recognitum, adauctum, et emendatum. Quibus accessere eiusdem authoris De hominis partu libri duo, nunc 3 ed.*, 2 vols. in 1 (Venice: apud Damianum Zenarium, 1602; cited hereafter as Augenio, *Epistolae* [1602]), vol. 1, bk. 11, 5 (to Pietro Pollastro), 135v: "Multa sunt, Petre, quae ad te amandum iam pridem impulerunt: . . . Alterum vero, quod ex conatus es patre, cui, ego plurimum debeo, quia cum olim Camerini cum Sebastiano Augenio fratre medicinam facerem, vix dici potest, quantum in me officiosus fuerit"; "faceret" is found in the place of "facerem" in idem, *Epistolarum medicinalium libri XII: Omnibus non medicis modo sed etiam bonarum literarum studiosis, admodum utiles* (Turin: apud heredes Nicolai Bevilaque, 1579; cited hereafter as Augenio, *Epistolae* (1579), 456).

4. Augenio named as his teachers Giovanni Argenterio (d. 1572), who taught at Pisa until 1555: "meus olim in gymnasio Pisano praeceptor" (Augenio, *Epistolae* [1602], vol. 2, bk. 4, *De consultandi . . . tractatio*, at 51r); Gianbattista Da Monte, who taught at Padua until his death in 1551: "Clarissimus praeceptor Ioannes Baptista Montanus (quem hic honoris caussa nomino)" (Augenio, *Epistolae* [1602], vol. 1, bk. 8, 1, 86v); and Giustiniano Finetti, who taught medicine at La Sapienza from 1539 until 1566, "Justinianus Finettus utriusque nostrum praeceptor" (Augenio, *Epistolae* [1602], vol. 2, bk. 2, [6] [to Pietro Crispo], 15r). On early modern travel for medical study, see Ole Peter Grell, Andrew Cunningham, and Jon Arrizabalaga, eds., *Centres of Medical Excellence? Medical Travel and Education in Europe, 1500–1789* (Farnham, Surrey: Ashgate, 2010).

5. The dates when Augenio' taught at Rome are uncertain. He is presumably to be identified with the Horatius de Monte Santo listed as a lecturer in logic in the university *rotulus* for 1559; see Emanuele Conte, ed., *I maestri della Sapienza di Roma dal 1514 al 1787: I rotuli e altre fonti*, 2 vols. (Rome: Istituto Storico Italiano per il Medio Evo, 1991), 1:30. As the index to the volumes of documents edited by Conte does not appear to contain any other entry that can be referred to Augenio, it seems likely that his medical teaching at La Sapienza may have taken place in one or more of the years 1560, 1562, 1564, and 1565, for which the *rotuli* are missing. Of his years of urban medical practice, Augenio stated, "Laboraveram sine intermissione annos viginti et octo medicinam faciens in praecipuis meae provinciae civitatibus" (Augenio, *Epistolae* [1602], vol. 2, bk. 5, [26], 70r). For the two works published in Camerino and Fermo, see idem, *Epistola, ad Hieronymum Cordellam medicum. In qua explicatur ratio curandi renes calculosos, & exulceratos. Conscripta pro eorum defensione, quae olim Cinguli ob ipso dicta sunt, contra opinionem Julii Cini Collensis medici* (Camerino: apud Antonium Gioiosum, [1575]), and idem, *Del*

modo di preservarsi dalla peste: libri tre, scritti volgarmente per beneficio commune (Fermo: appresso Astolfo de Grandi, 1577).

6. On the University of Turin under Emanuele Filiberto and his successor Carlo Emanuele, see Annamaria Catarinella, Irene Salsotto, and Andrea Merlotti, "Le istituzioni culturali," in *Storia di Torino*, vol. 3, *Dalla dominazione francese alla ricomposizione dello Stato (1536–1630)*, ed. Giuseppe Ricuperati (Turin: Giulio Einaudi editore, 1998), 523–67.

7. Augenio, *Epistolae* (1579); idem, *Epistolarum et consultationum medicinalium. Prioris tomi libri XII[–alterius tomi libri XII]*, 2 vols. in 1 (Venice: apud Damianum Zenarium, 1592); idem, *Epistolarum & consultationum medicinalium libri XXIIII in duos tomos distributi . . . Cum indice gemino: vno epistolarum; altero rerum & verborum copiosissimo. Quibus accessere eiusdem authoris, de hominis partu, libri II. Nunc primum in Germania ab innumeris mendis repurgati, ac in lucem emissi* (Frankfurt: apud heredes Andreae Wecheli, Claudium Marnium, & Ioannem Aubrium, 1597); idem, *Epistolarum medicinalium tomi tertii libri duodecim: in quibus non solum maximae difficultates ad medicinam et philosophiam pertinentes dilucidantur: sed etiam Alexandri Massariae Vicentini Additamentum apologeticum et disputationes secundum Hippocratis et Galeni doctrinam funditus evertuntur: nunc primum in lucem editi: cum duplici rerum indice copiosissimo* (Frankfurt: apud heredes Andreae Wecheli, Claudium Marnium, et Ioannem Aubrium, 1600; cited hereafter as Augenio, *Epistolae* [1600]); idem, *Epistolae* (1602); idem, *Epistolarum medicinalium tomi tertii libri duodecim: In quibus non solum maximae difficultates ad medicinam et philosophiam pertinentes dilucidantur; sed etiam Alexandri Massariae Vicentini additamentum apologeticum, et disputationem secundum Hippocratis et Galeni doctrinam funditus everturntur. Nunc primum Venetiis in lucem editi; cum duplici rerurm indice copiosissimo* (Venice: apud haeredem Damiani Zenarii, 1607).

For volumes 1 and 2, I cite from the edition of Venice (1602), giving volume, book, letter, and folio numbers. In this edition, in vol. 1 the letters are separately enumerated for each internal book, the number of each letter being included both in the table of contents and in headings of each item throughout the volume; in vol. 2, also divided into internal books, the numbering of individual items (which include some short treatises as well as letters) is found only in the table of contents and is consecutive throughout the volume. In citations of vol. 2, I include the item number from the table of contents in square brackets. I cite volume 3 from the edition of Frankfurt (1600) unless otherwise noted, giving the titles of the letters or short treatises as found in the body of the text (some of which have slight variations from the list of titles in the table of contents). I have also consulted the Turin (1579) edition of vol. 1 (any citations of this edition are by book, chapter, and page number), and the Venice (1592) edition of vols. 1 and 2. In the editions after 1579, beginning with 1592, a few extra letters were added to volume 1.

8. Orazio Augenio, *De sanguinis missione libri tres: in quibus non solum quid sit illud, quod vere indicat missionem sanguinis, & de plenitudine praeter recentiorum*

medicorum opinionem, disputatur: sed etiam maximae quaeque difficultates ad hoc negocium pertinentes, dilucidantur (Venice: [Comin da Trino], 1570); the original small work (16 cm high, containing seventy-four leaves, and divided into three internal books) was subsequently greatly expanded in subsequent folio editions: idem, *De ratione curandi per sanguinis missionem libri decem* (Venice: apud Damianum Zenarium, 1597), together with idem, *Disputationum de ratione curandi per sanguinis missionem ex Galeni sententia libri septem* (Venice: apud Damianum Zenarium, 1597); and idem, *De ratione curandi per sanguinis missionem libri XVII in duos tomos divisi: quorum prior decem, posterior septem continet* (Frankfurt: apud heredes Andreae Wecheli, Claudium Marnium, & Joan. Aubrium, 1598). Another edition of Augenio's work on bloodletting in ten books, which I have not seen, was published in Turin (1584). Idem, *Del modo di preservarsi dalla peste: libri tre, scritti volgarmente per beneficio commune* (Fermo: appresso Astolfo de Grandi, 1577), was subsequently published in a Latin version in Leipzig (1593 and 1598). Idem, *De febribus, febrium signis, symptomatibus, & prognostico libri septem, ab ipso authore ab anno 1568 usque ad 1572 singuli conscripti: nunc vero post ejus obitum ab Hilario Augenio authoris filio in lucem emissi* (Venice: apud haeredem Damiani Zenarii, 1607; first edition, 1604). Idem, *Quod homini certum non si nascendi tempus libri duo* (Venice: apud Joannem Baptistam Ciotum, 1595), is a slightly different version of his *De hominis partu libri duo*, included in vol. 2 of the Venice (1592 and 1602) editions of the *Epistolae*.

9. Paul Oskar Kristeller, *Iter Italicum: A Finding List of Uncatalogued or Incompletely Catalogued Humanistic Manuscripts of the Renaissance in Italian and Other Libraries*, 7 vols. (London: Warburg Institute, 1963–97): see 1:270, Mantova, Biblioteca Comunale, MS E III 28, "Horatius Augenius, lectiones medicae in scholis Taurini," and 2:402, Città del Vaticano, Biblioteca Apostolica Vaticana, Fondo Reginense Latino, MS 1271, "Horatius Augenius, commentary on Avicenna's *Canon*." I have not seen these manuscripts. Possibly the commentary in the Vatican represents Augenio's teaching at La Sapienza. On wine, see Augenio, *Epistolae* (1602), vol. 2, bk. 5 [27] (treatise in 14 chapters, dated 1590), 85r–92v: "Discipulis Taurinensibus. Quid vinum. Quae vini facultates. Qua proprietas. Quid proprie potus apud medicos, quid cibus, quid sit, quod nutritur. Quae partes ab ipso vino nutriantur, et qualis conveniat nutriendo modus." Augenio's interest in wine, its properties, and its medical uses (and when its use should be restricted) also shows up in a number other letters, especially Augenio, *Epistolae* (1602), vol. 2, bk. 2, [8] (on the use of wine in fevers), 21v–25v, and vol. 2, bk. 3, [19] (on the use of wine in catarrh and of a medicated wine for treating asthma, dated from Tolentino, 1576), 44r–45v.

10. Augenio, *Epistolae* (1600), vol. 3, letter of dedication (to Zacharia Contareno, Venetian senator and moderator of the University of Padua, dated 1598), *3r–v: "Sentio, Illustrissime Zacharia Contarene, me diutius fortasse, quam plerique

cuperent, adhuc libros meos de febribus comprimere, et in Galenum commentarios. Quam rem ne quis durius accipiat, ac secus quam par est interpretetur, ingenue fateor duas mihi potissimum rationes extitisse cunctationis meae. Altera est studium publicae utilitatis, cui quantum operae dandum sit, non te latet gymnasii huius moderatorem. Defensio meae dignitatis altera non in postremis habenda. Quem enim obsecro adeo robustae pietatis esse putas, qui si a quopiam laceretur, et in eius famam, quanta quanta [sic] colligi poterit involetur, non leviter saltem commoveatur? Alexander Massaria [1510–1598; professor of practical medicine at Padua from 1587] adeo me nullo meo merito acerbe pupugit, et labem meo nomini aspergere conatus est semel atque iterum, suo illo in Apologetico additamento, ut videar stylo subirato indulgere aliquid meo iure posse . . . Equidem sic existimo ab Alexandro Massaria multa esse disputata, quae non modo ad Hippocratis mentem, et Galeni medicinae oraculorum non sint, si quis rem suis momentis expenderit, sed tanquam illis contraria manifestem labem sint humano generi importatura. Cuius ergo sceleris mihi conscius essem, qua me faces die nocteque coquerent, Deus immortalis, si publicam rem, cui nostra maxime, ut auctor est Plato, desudat industria, tacitus in apertum sinerem adduci discrimen? . . . Hae caussa extiterunt, quam ob rem ego nunc minime promam, qua supra recensui lucubrationes meas sed pro illis Tertium hunc epistolarum Medicinalium Tomum emittam."

11. Augenio, *De febribus . . . libri septem.* The copy in the collection of the New York Academy of Medicine is bound with idem, *Epistolarum medicinalium tomi tertii libri duodecim* (Venice: apud haeredem Damiani Zenarii, 1607). Augenio's work on fevers was first published in Frankfurt (ex officina Matthiae Beckeri, sumptibus Jo. Theobaldi Schönwetteri, 1604) and also appeared in another Frankfurt edition the following year, both of which I cite from the online catalogue of the National Library of Medicine. For Augenio's desire not to anticipate Giovanni Argenterio's work on fevers, see idem, *Epistolae* (1602), vol. 2, bk. 3, [19], 44r (dated 1576 at 45v).

12. Augenio, *Epistolae* (1602), vol. 1, bk. 8, 1, 86r: "Accedit, quod prima pars meorum commentariorum in Galeni libros, quos aedere propediem constituo, adeo tenent me dies noctesque implicitum[.]"

13. Augenio, *Epistolae* (1579), *3r: "Nam si quod mihi vita superstite ocium ad studia dabitur (ut aliquando dari abste, et a Serenissimo Patre opto, et confido) secundam, atque tertiam epistolarum et responsorum medicinalium partem, quae nunc transcribitur, aliquemque graviora, praesertim libros de Febribus, et de Placitis Averrois et Galeni iam prope absolutos sub tuis pariter auspiciis in lucem profferre tentabo."

14. Augenio, *Epistolae* (1579), 1.1, pp. 1–19 (dated at end 1558): "Ludovico Augenio patri et medico clarissimo, Horatius Augenius filius S.P.D. epistola prima. Agitur de ptissana [sic], et eius usu." At p. 13, in margin: "Ioannis Manardi opinio de ptissana, libr. 5, epist. 2"; at p. 15, in margin: "Augenii opinio contra Manardum."

15. Augenio, *Epistolae* (1602), vol. 2, bk. 3, [22], 48v, in margin: "Historiae admirandae de pica lege Langii epistolas"; for the anecdote in question, see Lange, *Epistolae* (1605), 2.12, pp. 548–50 (= *Miscellanea* 1560, pp. 65–66]). Augenio, *Epistolae* (1602), vol. 2, bk. 6, [37], 102v, citing the recipe in Lange's bk. 3, letter 1, chapter 11, which was presumably from Johannes Lange, *Epistolarum medicinalium volumen tripartitum, denuo recognitum, et dimidia sui parte auctum* (Frankfurt: apud heredes Andreae Wecheli, Claudium Marnium et Ioannem Aubrium, 1589), p. 856[misprint for 956].

16. The letters to family members are in Augenio, *Epistolae* (1602), vol. 1: to Lodovico Augenio (father), bk. 1, 1 and 2 (both dated 1558), 1r–16r; to Giulio Picchino (son-in-law's father), bk. 3, 3, 38v–40v, and bk. 7, 1–8 (several of the letters are dated 1570 or 1572), 75r–85v; to Fabrizio Augenio (brother), bk. 12, 1 (dated 1574) and 2, 139r–140v; to Bernardino Montarino (nephew), bk. 12, 3, 140v–141v; and Augenio, *Epistolae* (1602), vol. 2: to Giulio Picchino, bk. 1, [3], 10r–11r; to Lodovico Picchino (son-in-law), bk. 6, [28], [29] (dated 1581), [36], 93r–95v, 102r–v; bk. 7, [45], 112v–116v; bk. 8, [51], 123v–126v; bk. 11, [68], 155v–157r; to Sebastiano Augenio, called Paparella (cousin), bk. 9, [54], 130r–131v. The last-named individual was the author of several medical treatises (listed in the online catalogue of the National Library of Medicine under the author name "Sebastiano Paparella").

17. Augenio, *Epistolae* (1602), vol. 2, bk. 4, [23] (to Antonio Lobetto, *medicus* of the Duke of Savoy), 50v–64v: "De consultandi, sive collegiandi (ut vocant) ratione, qua medici passim utuntur tractatio." Augenio, *Epistolae* (1600), vol. 3, bk. 2, 5 (to Alessandro Massaria), pp. 34–62: "De medicis novatoribus disputatio."

18. Augenio, *Epistolae* (1602), vol. 2, bk. 2, [8] (to Pietro Crispo), 32r: "Non sunt admodum stabiles, quae sermone solo disputationes instituuntur: nam aut non bene auditum, aut non dictum, aut linguae transcursum adfuisse, dicere plerique solent. At certe semper manet scriptura. Praetermitto hac de caussa, quod mecum disputans, dixisti; vel ut rectius fortassis loquar, dicere voluisti; atque hisce litteris scribo quid sentiam, ut, si non placuerint tibi, audacter possis, quod te facturam pollicitus es disputare mecum." Pietro Crispo taught *medicina theorica* at La Sapienza from 1567 to 1576; see G. M Carafa, *De Gymnasio Romano et de eius professoribus libri duo* (Bologna: Forni, 1971; originally published Rome, 1751), 2:357, and Conte, *I maestri della Sapienza di Roma*, 1:46–112 (*rotuli* for 1567 to 1576) and 2:961, s.v. "Petrus Crispus."

19. For example (in this particular case, it was Augenio who was the recipient of a letter originally sent to someone else), Augenio, *Epistolae* (1602), vol. 1, bk. 8, 3 (to Giulio Bossello), 94r: "Plinius Melus nostrum utriusque amicissimus, tuas ad illum legendas mihi obtulit litteras: in quibus docte admodum de affectionibus praeternaturam coniugis illius edisseris."

20. Augenio, *Epistolae* (1602), vol. 2, bk. 3, [18] (to Francesco Cirocco, from Cingoli, 1571), 41r: "Scripsi tractationem diebus praeteritis de cauteriorum, et sina-

pismatum usu, in catarrhi potissimum curatione, observando, quia medicos noscebam quosdam, generosis hisce remediis abuti: eam edere sub tuo nomine placuit, non, ut tu ipse adisceres aliquid, cum iam totius artis palmam sis consecutus: sed, ut memineris, meminisse me tibi debere, quod iam diu promiseram."

21. The period noted for Augenio's practice at Cingoli is based on dates given in his letters: more than a dozen letters in Augenio, *Epistolae* (1602), vols. 1 and 2, are dated from Cingoli between 1570 and 1574 (in the printed collections of Augenio's *Epistolae* some, but by no means all, letters conclude with the date and place of writing). But the earliest dated letter from Augenio's years as a town practitioner comes from Cingoli, dated 1562 (Augenio, *Epistolae* [1602], vol. 1, bk. 3, 2 [with date at end], 34r–38v), presumably representing an earlier period in, or visit to, that town.

22. For the letters to Sanzio, see Augenio, *Epistolae* (1602), vol. 2: bk. 4, [24] (addressed to "Ioanni Francisco Sanctio Cingulano, Rochae contradae medico, discipulo amantissimo," from Turin, 1580), 65r–68v; bk. 6, [33] (dated 1582), 97v–98v (dated from Turin, 1581); bk. 7, [40] (dated from Cingoli, 1570), [41–43], 106r–111v; bk. 8, [49] (dated from Cingoli, 1574), [50], 120r–123v. The letter to a colleague, in Augenio, *Epistolae* (1602), vol. 2, bk. 6, [34] (to Francesco Cirocco), begins (at 98v), "Quod, in Illustrissimi Cardinalis Symposio cum Sanctio meo, disserendo, non potui: id nunc, ut tibi satisfaciam, absolvere conabor" and ends (at 101r), "Iam igitur, vir clarissime audivisti quid sentiam de maximis difficultatibus, quas proposuit mihi ingeniosissimus Ioannes Franciscus Sanctius."

23. Augenio, *Epistolae* (1602), vol. 2, bk. 7, [50], 123v: "Caeterum, dulcissime Sanctii, hoc remedium periculosum est duplici nomine, primum quidem ob quam habere certitudinem poterit, ne ultra vacuatio progrediatur? Atque licet adstringentia postea simus exhibituri, tamen quis confidet medicamento validissime purgante utilitatem aliquapiam allatura? Sic frequenter non modo accidit animi defectus, sed etiam sincopes, et mors. Secus res habet in venae sectione, hanc ipsam cum volumus sistemus imposito digito, at non ita in purgatione contingere certissimum est . . . Idcirco nunquam observasti me in hisce montibus hac usum fuisse purgandi ratione, neque posthac etiam sum usurus. Admoneoque te ipsum, ut omnino abstineas in hac praesertim aetate iuvenili. Nam si aliquid extra spem contingeret tibi, non vacares apud vulgum crimine internecionis. Oportet certe maxime cautum esse in factitanda medicina, ac omnium maxime iuvenem."

24. Iain M. Lonie, "Fever Pathology in the Sixteenth Century: Tradition and Innovation," in *Theories of Fever from Antiquity to the Enlightenment*, ed. W. F. Bynum and Vivian Nutton, Medical History Supplement No. 1 (London: Wellcome Institute for the History of Medicine, 1981), 25n22.

25. Augenio, *Epistolae* (1602), vol. 1, bk. 10, 5 (dated from Tolentino, 1576), 124v: "Ioannes Fernelius recentiorum medicorum citra controversiam princeps." On Fernel, see Hiro Hirai, "Ficin, Fernel et Fracastor autor du concept de semence:

Aspects platoniciens de *seminaria*," in *Girolamo Fracastoro fra medicina, filosofia e scienze della natura*, ed. Alessandro Pastore and Enrico Peruzzi (Florence: Olschki, 2006), 245–60; James J. Bono, *The Word of God and the Languages of Man* (Madison: University of Wisconsin Press, 1995); and Linda Deer Richardson, "The Generation of Disease: Occult Causes and Diseases of the Total Substance," in *The Medical Renaissance of the Sixteenth Century*, ed. A. Wear et al. (Cambridge: Cambridge University Press, 1985), 175–94.

26. Augenio, *Epistolae* (1602), vol. 1, bk. 10, 3, 119v: "non sunt autem usi medicamentis nostri temporis, quia his caruere: in qua re nostra aetas antiquitatem superat."

27. For example, Augenio, *Epistolae* (1602), vol. 2, bk. 1, [4], 12v: "Dixi de hoc argumento plura in libro de concoctione"; and vol. 2, bk. 2, [12], 32r: "Sunt haec, Fauste, libro tertio meae tractationis de purgandi ratione diligenter, nisi fallor, explanata." As far as I know, these works—if completed—were never printed; I am not aware of manuscripts.

28. Augenio, *Epistolae* (1602), vol. 2, bk. 5, [26], 70v: "Verum cum primum tomum epistolarum medicinalium adversus importunitatem cinicam edere in lucem statuissem, variasque colligissem epistolas, forte at manus tuae illa pervenerunt litterae, et tunc haud amplius cunctandum ratus, responsionem dedi, eamque cum aliis epistolis publicavi." But on another occasion, Augenio claimed his letters had been collected and published by his son-in-law without his knowledge; perhaps he was referring to vol. 2 of his *Epistolae*, though, as the remark occurs early in vol. 2, this does not seem likely. See Augenio, *Epistolae* (1602), vol. 2, bk. 3, [15] (dated from Turin, 1589), 37r: "Medicinales Epistolas olim meo genero, et me inscio editas." Augenio's letters concerning Cini's quarrel with Cibo over the death of Monaldus Clementinus are in Augenio, *Epistolae* (1602), vol. 1, bk. 6, 1 (to Giulio Cini, dated from Cingoli, 1572), 62r–72v; and vol. 1, bk. 9, 2 (to Francesco Cirocco), 101v–106v. The works on kidney complaints are idem, *Epistola, ad Hieronymum Cordellam medicum* (published in 1575), and Giulio Cini, *Apologia adversus Horatium Augenium Picoentium et eius sequaces* (Perugia: apud Balthassarem Salvianum, 1576). In another letter to Cordella (Augenio, *Epistolae* [1602], vol. 1, bk. 4, 1, 41r–46v), Augenio thanked Cordella for support against his critics. Franceschini is referred to elsewhere as a cleric for whom Augenio prescribed medical treatment in consultation with Cordella (Augenio, *Epistolae* [1602], vol. 2, bk. 12, [75] [to Francesco Cirocco], 169v).

29. Augenio, *Epistolae* (1602), vol. 1, bk. 7, 2, 4–6, 8 (all to Giulio Picchino), 76v–77r, 79v–82v, 84v–85v, concern grains (used for medicinal plasters as well as in *ptisana*) and honeyed wine. Augenio, *Epistolae* (1602), vol. 1, bk. 1, 1 (to his father, Lodovico Augenio), also concerns *ptisana*. One other early letter relating to natural history is Augenio, *Epistolae* (1602), vol. 2, bk. 6, [30] (to Egidio Franceschini of Cingoli, dated from Cingoli, 1571), 95v–96r, crediting Franceschini with identifying the true cave sparrow known to the ancients, "quoniam vero reparationis

praestantissimi huiusce medicamenti tu ipse author extitisti," and arguing that the bird known as "cauda tremula" was not the same as the "passerculus troglodytes."

30. Augenio, *Epistolae* (1602), vol. 2, bk. 6, [34], 99v.

31. On Lodovico or Luigi Augenio, see Gaetano Marini, *Degli archiatri pontifici volume primo, nel quale sono i supplimenti e le correzioni all'opera del Mandosio* (Rome, 1784), 342; Panelli d'Acquaviva, *Memorie*, 103–4. For a very comprehensive study of medicine in sixteenth-century Rome, see Elisa Andretta, *Roma medica: Anatomie d'un système médical au XVIe siècle* (Rome: École française de Rome, 2011).

32. Augenio, *Epistolae* (1602), vol. 2, bk. 5, [26], 70v: "novus ad dicendum homo non est, qui multo antea Romae publice docuit."

33. Augenio, *Epistolae* (1602), vol. 1, bk. 12, 1, 139r: "Semper tuus ab urbe discessus mihi displicuit: ita nunc tuus ad eas gentes accessus affecit me dolore maximo." The body of the letter offers advice for Fabrizio's urinary problem, including (at 140r): "aquam hanc [sc. Tyberinam], ubi optime fuerit expurgata (de hac nam loquendum) non modo facere calculos non puto, sed ab illis preservare arbitror: duae sunt coniecturae, quae id mihi persuadent: una est, quod aqua haec copiose etiam bibita nullum in praecordiis vitium relinquit, sed ob levitatem subtilitatemque substantiae facile penetrat . . . Adde his quod Paulus Tertius Pont. Max. foelicis memoriae, ut in rebus omnibus erat diligentissimus, sic in victus ratione alios suae aetatis homines, ut fama est, superavit: quocunque ibat secum hanc deferri aquam curabat: at vero cum senes propemodum omnes calculo sint maxime obnoxii, mirum fuisset, si ex aqua Tyberina in hunc ille afffectum non incidisset." On the midcentury Roman controversy about the wholesomeness of Tiber water, see Nancy G. Siraisi, "*Historiae*, Natural History, Roman Antiquity and Some Roman Physicians," in *Historia: Empiricism and Erudition in Early Modern Europe*, ed. Gianna Pomata and Nancy G. Siraisi (Cambridge, MA: MIT Press, 2005), 325–26, and the bibliography cited therein; for the anecdote about Paul III, see Nancy G. Siraisi, *History, Medicine, and the Traditions of Renaissance Learning* (Ann Arbor: University of Michigan Press, 2007), 95, and the bibliography cited therein.

34. Augenio, *Epistolae* (1602), vol. 2, bk. 2, [6], 18v, in margin: "Lege Eustachii librum de Vena sine pari"; vol. 2, bk. 9, [59], 134v, claims that Lodovico Augenio and "Bartholomaeus Eustachius Picenus" were at one time both practicing medicine in Pesaro in Le Marche, and that they attended a patient there together. For the origins and career of Eustachi (d. 1574), see Maria Mucillo, "Eustachi (Eustachio, Eustachius), Bartolomeo," *DBI*, www.treccani.it/enciclopedia/bartolomeo -eustachi_(Dizionario-Biografico)/. Augenio, *Epistolae* (1602), vol. 2, bk. 3, [17] (dated 1569), 41r: "Andreas Baccius Elpidianus amicus meus, qui nunc Romae medicinam facit." Antonio Porto of Fermo is the addressee of Augenio, *Epistolae* (1602), vol. 2, bk. 7, [44], 111v–112r. Bacci and Porto were both among the physicians of Sixtus V (reigned 1585–90), according to Marini, *Degli archiatri pontifici*,

1: 462–64; as the heading of Augenio's letter to Porto terms him "Sixti V Pontificis Maximi a Cubiculo Medico" (at 111v), the letter was presumably written after 1585. On Porto, see also Luigi Belloni, "L'aneurisma di S. Filippo Neri nella relazione di Antonio Porto," *Rendiconti del'Istituto Lombardo di Scienze e Lettere: Classe di Scienze Matematiche e Naturali* 83 [vol. 14 of ser. 3] (1950): 665–80.

35. On Cordella, who came from Fermo and was another of Filippo Neri's doctors, see Panelli d'Acquaviva, *Memorie*, 207–9, and Marini, *Degli archiatri pontifici*, 1:476–77; in addition to being the addressee of Augenio, *Epistolae* (1602), vol. 1, bk. 4, 1, 41r–46v, Cordella is mentioned respectfully in Augenio, *Epistolae* (1602), vol. 1, bk. 7, 5 (dated 1570), 81v; vol. 1, bk. 9, 2, 102v; vol. 2, bk. 8, [47] (a *consilium*), 118r: "Hieronymus Cordella medicus absolutissimus, cuius doctrinae, diligentiae, atque observationi ita plane confido"; and elsewhere. An anecdote in Augenio, *Epistolae* (1602), vol. 1, bk. 8, 1, 88r, illustrating the point that physicians should be careful about giving patients up for dead, also suggests a connection, or solidarity, among Roman residents of the same regional origin: "Anconitanum vidi Romae sepultum, qui ad aperto post mensem sepulcro sedens mortuus, manus habens circa caput, inventus est."

36. Augenio, *Epistolae* (1602), vol. 2, bk. 1, [6] (to Pietro Crispo, dated from Rome, 1568), 15r–19v; vol. 2, bk. 2, [13] (to Pietro Crispo), 32r–33v; vol. 2, bk. 7, [39] (to Alessandro Petroni, dated from Cingoli, 1570), 104v–106r. On Petroni (d. ca. 1585), who was one of the physicians of Gregory XIII, see Marini, *Degli archiatri pontifici* 1:454–55, and Siraisi, *History, Medicine*, 168–69, 172–73, 181–82.

37. Augenio dated letters from Rome during return visits in 1568 and 1575: Augenio, *Epistolae* (1602), vol. 2, bk. 2, [6] (to Pietro Crispo), 15r–19v (1568); vol. 2, bk. 11, [66–67] (to Giuseppe Favorino de Clavari), 150v–155v (1575). The 1575 trip is referred to in the letter to Fabrizio Augenio in Augenio, *Epistolae* (1602), vol. 1, bk. 12, 1 (1574), 139r–140v, at 140r–v: "Reliqua . . . quamprimum sequenti anno Iubilei ad urbem redieris, et nos viseris, absolventes exequemur." Augenio, *Epistolae* (1602), vol. 1, bk. 10, 5 (dated from Tolentino, 1576), 123r, appears to refer to another brief visit to Rome in 1576. Augenio, *Epistolae* (1602), vol. 1, bk. 1, 1 and 2 (to his father Lodovico Augenio, both dated from Rome, 1558), 1r–16r, were presumably written when Augenio was studying or teaching in the city.

38. Augenio, *Epistolae* (1602), vol. 2, bk. 1, [1], 2v: "Dum olim Romae profiterer medicinam, orta fuit discordia inter Bartholomaeum Eustachium Sanctoseverinatem, et Realdum Columbum, anatomicos quidem, ut eorum testantur monumenta, praestantissimos, negante illo, hoc autem affirmante adesse interseptum virginale."

39. Augenio, *Epistolae* (1602), vol. 1, bk. 4, 2 (to the surgeon Durante Scacchi, dated from Tolentino, 1576), 46v–50r. At 48r: "hic ego tacitus praetermittere non possum, quod observavi Romae in filio cuiusdam Veneti, qui Typographus erat Pontificius calculum in vesica patiebatur, et ab omnibus deploratus, de sectione

cogitabatur, et iam pater cum Nursino artifice convenerat: quando Reverendus quidem Sacerdos, et si bene recordor, ex societate Iesu proposito medicamento quodam insperato, adolescentem sanavit . . . Testis huius rei est Iuvenalis Ancinas Fossanensis medicus clarissimus." At 50r: "Secundo loco, non accedat artifex temere ad hoc artificium . . . Postremum opus est, ut qui egregiam hanc effecturus est actionem, calleat artem secandi corpora, anatomen Graeci vocant: quia turpis est certe ac detestanda nostri temporis consuetudo, ut homines rudes ad artificium hoc perdifficile admittantur . . . Hinc sit, Scacche, ut eorum, qui secantur ab istis circumforaneis medicis maior pars pereat, qui vero evadunt aut fistulas perpetuas portant, aut generandi potestam laesis vasis spermaticis amittunt. Infoelicitatem nostrorum temporum misertus omnipotens Deus nobis dedit egregium Varolum [*sic*] Bononiensem, virum quidem non modo in omnibus artis operibus clarum, sed in hoc certe clarissimum: quandoquidem omnes fere, quibus ille manus adhibebat, sanabantur, ut Romae vidi." Augenio, *Epistolae* (1602), vol. 2, bk. 9, [59] (to Sigismund Kolreuter, dated from Turin, 1582), 134v–135r, at 134v: "Casus fuit. Filium Zanetti Pontificii tipographi Romae expurgandum susceperamus doctissimus D. Juvenalis Ancinas Fossanensis et ego." At 135r: "Hanc rationem secandi mea persuasione adhibebat Romae quidam ex meis discipulis, ac omnes fere sanitati restituebat." For the arrival in Rome of Ancina and Varolio, see C. H. Bowden, *The Life of B. John Juvenal Ancina* (London, 1891), 9; Charles D. O'Malley, "Varolio, Costanzo," in *Dictionary of Scientific Biography*, ed. Charles C. Gillispie, 16 vols. (New York: Charles Scribner's Sons, 1970–80), 13:587.

40. Augenio, *Epistolae* (1602), vol. 2, bk. 7, [39] (to Alessandro Petroni, dated from Cingoli, 1570), 104v–106r. Ibid., vol. 1, bk. 9, 2 [misnumbered, actually letter 3], 107r–109v, answers the request of the vicar general of the diocese of Osimo. Ibid., vol. 2 , bk. 6, [38] (to Cardinal Bozzuto), 103v–104r ; see also idem, *De hominis partu libri duo* (printed with Augenio, *Epistolae* [1602]), bk. 2, chapters 10–16, 25r–29r, especially chapter 16, "Octimestrem partum nasci aliquando vitalem." On Bozzuto, who was named cardinal in May of 1565 and died in October of that year, see Roberto Zapperi, "Bozzuto, Annibale," *DBI*, www.treccani.it/enciclopedia/annibale -bozzuto_(Dizionario-Biografico). Orazio Augenio, *De hominis partu libri duo*, included in the 1597 and 1602 editions of idem, *Epistolae medicinales*, was originally published in a slightly different version as idem, *Quod homini certum non sit nascendi tempus. Libri duo* . . . (Venice: apud Ioannem Baptistam Ciotum, 1595); Augenio seems to have begun, or at any rate planned, the treatise before 1579, to judge from Augenio, *Epistolae* (1579), 8.2, pp. 301–13, at 301: "constitui dare tibi literas, ut tu ipse ex his cognoscas, Valteri, me sane habere nihil antiquius, quam tueri veritatem ipsam: quod me fecisse iudicabunt ii, qui libros meos de homines procreatione, ac partu perlegent." This letter (= Augenio, *Epistolae* [1602], vol. 1, bk. 8, 2, 90v– 94r) contains the only substantial discussion of astrology in Augenio's *Epistolae*, mostly in connection with the defense of his views on the survivability of the

eighth-month fetus. Concetta Pennuto has in preparation a study of the history of the belief that survival was impossible for a fetus born in the eighth month of pregnancy.

41. Augenio, *Epistolae* (1602), vol. 2, bk. 1, [2] (dated from Turin, 1585), 8r–10r: "Senatui Pedemontano." Augenio, *Epistolae* (1600), vol. 3, bk. 9, 22, p. 367: "Casus propositus sacro et venerando collegio Patavino de suspicione exhibiti veneni, a praefecto urbis Romae [the official in charge of criminal justice in Rome]" and 23, pp. 367–68: "Responsum Horatii Augenii et Aemilii Campilongi, pro sacro et amplissimo collegio Patavino circa propositum casum" (dated Padua, 1598). On the establishment of the Piedmont Senate by Emanuele Filiberto in 1560 and its early composition and responsibilities, see Carlo Dionisotti, *Storia della Magistratura Piemontese*, 2 vols. (Turin, 1881), 1:99–102.

42. Augenio, *Epistolae* (1602), vol. 2, bk. 9, [53] (dated at end Turin, 1585), 129r–130r: "Illustrissimis, Ozascho magno cancellario Serenissimi ducis Sabaudiae, et Ludovico Puteo primario Praesidi Pedemontani Senatus. Ostenditur neque sal, neque oleum olivarum contagionem recipere pestilentem posse." At 129v: "Idem etiam facio iudicium, de ea quaestione, an sal possit venenari a scelestis hominibus, cum enim id mihi sit incognitum, aliquid certi pronunciare non debeo." The date on the letter is problematic, as—according to Dionisotti, *Storia della Magistratura Piemontese*, 2:197–98 and 244—Cacherano Malabaila d'Osasco died in 1580 and Lodovico Dalpozzo died in 1582.

43. Augenio, *Epistolae* (1602), vol. 2, bk. 9, [52], 127r–129r: "Senatui Pedemontano. Quod medicamenta componere non deroget nobilitati artis. Pro Domino Iulio Contarino medico." See David Gentilcore, *Medical Charlatanism in Early Modern Italy* (Oxford: Oxford University Press, 2006), especially 216–26.

44. Augenio, *Epistolae* (1602), vol. 2, bk. 9, [52], 127v: "In praecipuis Italiae civitatibus lege cautum est, pharmacopolas medicamenta expurgantia, et magnas confectiones (ita quod vocant theriacam, mitridatum, auream alexandrinam, ac eius generis alias) componere non posse sine assistentia unius ex collegio deputati, aut plurium medicorum. Neque hoc est cuiusvis medici officium, sed eius in primis, qui fuerit ea in arte exercitatus: atque ita humano generi optime consulitur. At in iis civitatibus, in quibus haec consuetudo non adest, periculosum videtur mihi, omnia in potestate pharmacopola relinquere." On the supervisory responsibilities of the physician appointed to the position of *protomedico* and the College of Physicians of Rome, see David Gentilcore, "'All that Pertains to Medicine': *Protomedici* and *Protomedicati* in Early Modern Italy," *Medical History* 38 (1994): 121–42, and Fausto Garofalo, *Quattro secoli di vita del Protomedicato e del Collegio de Medici di Roma: Regesto dei documenti dal 1471 al 1870*. Pubblicazioni dell'Istituto di storia della medicina dell'Università di Roma, Collezione C: Studi e ricerche storico-mediche (Rome: Istituto di storia della medicina dell'Università di Roma, 1950). The recent medical and pharmacological authors cited in Augenio's report

(*Epistolae* [1602], vol. 2, bk. 9, [52], 128r) are Pietro Andrea Mattioli, Johann Crato, Giovanni Cortusio, and Prospero Bogaruccio.

45. Augenio, *Epistolae* (1602), vol. 2, bk. 6, [32], 96v–97v.

46. Augenio, *Epistolae* (1600), vol. 3, bk. 8, 18 (treatise in fourteen chapters), pp. 288–321: "Theodoro Zwingero Basiliensi medico et philosopho excellentissimo et amico singulari. Quod pueris ante annum decimum quartum tuto liceat interdum mittere sanguinem, adversus Alexandrum Massariam" (title from "Index epistolarum" at the beginning of the volume). In addition to the promised discussion of bloodletting, the letter includes the remark (at 288): "De Theologicis illis clarissimi viri patris tui scriptis pergratam mihi feceris, si a te illa habuero. Neque video cur non liceat." The online catalogue of manuscripts of the Universitätsbibliothek, Basel, includes five other letters from Augenio to Theodor Zwinger, dated between 1580 and 1587 (Basel, Universitätsbibliothek, *Katalog Handschriften und Nachlässe*, G2 II.8, 040; Frey-Gryn I.15, 17; Frey-Gryn II.8, 64–66; I have not seen these manuscripts).

47. Augenio, *Epistolae* (1602), vol. 2, bk. 7, [44], 111v–112r: "Antonio Porto Firmano Sixti V Pontificis Maximi a cubiculo medico. De ferina tussis ab humoribus falsis et crassis, cum febre continua lenta, curatione. Responsum pro Reverendissimo Iulio Ottinello Episcopo Fanensi, et Nuncio apud Serenissimum Carolum Emanuelem Sabaudiae Ducem" (Ottinello was the papal nuncio at Turin, 1589–93). Augenio, *Epistolae* (1602), vol. 2, bk. 3, [17] (to Girolamo Vitelleschi, "primario Anconae medico," dated from Osimo, 1569), 39r–41r, describes a consultation between Augenio and Vitelleschi about the treatment of a bishop. For the apostolic secretary Egidio Franceschini, see Augenio, *Epistolae* (1602), vol. 2, bk. 12, [75] (to Francesco Cirocco), 169v; vol. 2, bk. 6, [30] (to Egidio Franceschini, dated from Cingoli, 1571), 95v–96r. Augenio, *Epistolae* (1602), vol. 1, bk. 9, 3 [actually letter 4], 110r–111v: "R. P. Franciscano Pennensi theologo insigni, et amico optimo." Augenio, *Epistolae* (1600), vol. 3, bk. 10, 27, pp. 394–95: "Historia monialis Genuensis sanguinem expuentis missa ad Ludovicum Mercatum Hispanum Medicum Philippi secundi Hispaniarum regis, et ad Horatium Augenium medicinae in Patavino gymnasio professorum primum" (title from "Index epistolarum" at the beginning of the volume); vol. 3, bk. 10, 28 (dated from Toledo, 1596), p. 396: "Responsum Ludovici Mercati Philippi magni regis Hispaniarum protophysici"; vol. 3, bk. 10, 29, pp. 396–97: "Ad superiores literas Horatii Augenii responsum de sputo sanguinis." Controversy over the treatment of this last case was sufficient to lead to the publication of *Risposta di Hieronimo Veneroso nobile Genovese alla querela sotto nome di difesa intorno allo sputo di sanguine* (Florence: Vittorio Baldini, 1597), which records the differing opinions of physicians, Augenio among them; therein the patient is identified as Costanza Giustiniano.

48. Augenio, *Epistolae* (1602), vol. 2, bk. 3, [16], 37v–39r: "Responsum in caussa dissolutionis matrimonii inter Dominum, et Dominam. Ad Auditores Romanae

Rotae." At 38v: "fides non est adhibenda mulieri, quae virum suum odio prosequi-
tur immortali: fuit autem odii caussa, quod eam alapis percusserit, ut in actis ap-
paret: tuncque illa suum patrem fugiente, factum fuit divortium apparetque
suum obiurgasse virum, minatumque fuisse, amplius ad suam domum non esse
reversuram . . . Dicent, in congressu cum sponsa non potuit erigere . . . Repeto,
ex eo quod quispiam aliquando non erigat, non tollitur erigendi potentia alio
tempore . . . Praeterea manifestae adfuerunt caussae a medicis conscriptae, impedi-
entes erectionem, odium scilicet, commemoratio rerum praeteritarum, quae vel
maxime ad illius honorem pertinebant. Rixa, tumultus cum socrum, et socero, et
verecundia. Sciebat enim pro foribus assistere obstetricem, archiepiscopalem no-
tarium, quatuor medicos, ut magnam aliorum turbam praetermittam."

49. Augenio, *Epistolae* (1602), vol. 2, bk. 10, [62] (dated from Turin, 1588),
141v–148r: "In causa dissolutionis matrimonii inter Dominum, et Dominam, Augenii responsum. Multa explanantur de naturali potentia, et impotentia ad co-
itum, de paralesi virgae, de maleficiatis, ac frigidis, atque sigillatim caussa omnes
recensentur, quae impotentem aliquem reddant." Incipit: "Quod ex tempore in
praestantissimorum hominum corona olim a me dictum fuit, id nunc ipsum, ma-
turiori facta consideratione, scribendum duco." The account of the couple's problems leaves little doubt that this is the same case as in the earlier report to the
Auditores Romanae Rotae. The new emphasis on witchcraft appears at 144r–v
(chapter 5): "Maximam adfuisse in proposito casu maleficii suspicionem." On the
history of medieval beliefs about magic and impotence; their presence in medieval
learned writing on theology, canon law, and medicine; and their transformation in
the context of early modern witchcraft beliefs, see Catherine Rider, *Magic and
Impotence in the Middle Ages* (Oxford: Oxford University Press, 2006). On impotence in medieval canon law, see James A. Brundage, *Law, Sex, and Christian Society in Medieval Europe* (Chicago: University of Chicago Press, 1987), 290–92,
377–78, 456–58.

50. Augenio, *Epistolae* (1602), vol. 2, bk. 1, [1] (treatise in thirteen chapters,
dated from Turin, 1587), 1r–8r: "Reverendissimo in Christo Patri Domino Domino [*sic*] Seraphino Olivario, S.D.N. PP. Capellano, eius Sacri Palatii Caussarum
Auditori, Domino suo colendissimo. Ostenditur virgines foeminas eam non habere
ex natura membranam, quam nonnulli interseptum, alii claustrum virginale, alii
hymen vocant. Nullum certum, propriamque signum esse virginitatis. Et quod olim
medicarum mulierum, quas hodie obstetrices vocant, ex Platone, et Galeno extit-
erit officium"; Séraphin Olivier-Razali (1538–1609) became an auditor of the Roman Rota in 1564. At 1v: "Virgines non habent aliquam corporis partem, qua sint
destitutae mulieres. Deinde. Nullo certo signo distingui possunt virgines a muli-
eribus. Haec mea fuit responsio, a Taurinensibus medicis, in omni scientiarum
genere praestantissimis, (ut dixi) confirmata: quam responsionem tu ipse temerar-
iam iudicasti." At cap. 9, 7r: "Quod probabile est, tum verum, tum falsum esse . . .

Quandoquidem certissimum est, virgines aliquando incidere in hunc affectum, cum praeternaturam sese habent: at vero secundum naturam habere ipsum interseptum virginale esse omnino impossibile profiteor." On the range of Renaissance views on the hymen, investigations of female genitalia by sixteenth-century anatomists, and the belief in the universality of bleeding at first intercourse, see Valeria Finucci, "Devianza sessuale e imperativi genealogici: Il caso di Margherita Gonzaga," *Acta Histriae* 15 (2007): 385–98, at 385–90.

51. Augenio, *Epistolae* (1602), vol. 2, bk. 1, [1], cap. 12, 8r: "Quibus obstetricibus fides adhibenda sit . . . Deinde addo eiusmodi non reperiri obstetrices nostra aetate, praesertim vero in Italia, at omnium minime in hac Pedemontana regione, in qua rudes quaedam mulierculae omnium rerum ignarissimae hanc faciunt artem. Deberent saltem incumbere dissectioni corporum, aut huius tantum habere notitiam, quantam illarum postulat officium. Propterea Hispani cogunt obstetrices interesse, cum secantur corpora, quae res ubique deberet observari." For Joubert's discussion, see Laurent Joubert, *Popular Errors*, trans. and annotated by Gregory David de Rocher (Tuscaloosa: University of Alabama Press, 1989), book 5, chapter 4, 211–21. On the dates of the sixteenth-century translations of Joubert's work, see De Rocher, "Introduction," in Joubert, *Popular Errors*, xx. For Augenio's praise of Joubert, see Augenio, *Epistolae* (1600), vol. 3, bk. 2, 5, (to Alessandro Massaria), pp. 34–62: "De medicis novatoribus disputatio," in cap. 2, p. 36.

52. Augenio, *Epistolae* (1602), vol. 2, bk. 1, [1], cap. 7, 6r: "Inspectionem mulierum, quae admittitur de Iure Pontificio, signum fallax existere ipsius virginitatis"; ibid., cap. 8, 6v: "Augenii responsum non fuisse contrarium Iuri Pontificio et rationis Olivarii contra Augenium explicatio."

53. Augenio, *Epistolae* (1602), vol. 1, bk. 3, 1 (treatise in thirteen chapters, dated from Tolentino, 1576), 16v–24v: "Hieronymo Capivaccio medico clarissimo. Cinica subvertitur Apologia, in qua dicuntur quamplura de natura calculi, eiusque caussis; de plano ulcere, an compositus sit morbus; de indicationibus ad calculum extirpandum; de usu lactis asinini, et balnei aquae dulcis, pro eorum confirmatione, quae ab Augenio scripta sunt libro de calculosis, et ulceratis renibus."

54. Augenio, *Epistolae* (1579), *2r–*3v: "Carolo Emanueli a Sabaudia Serenissimo Pedemontium principi Horatius Augenius foelicitatem." At *2v: "Cum enim defuncto Francisco Valleriola viro clarissimo, medicinamque primo luce profitenti, almi huius gymnasii Moderatores prudentissimi, quem substituerent ex variis regionibus conquirerent, visus sum illis idoneus ad id muneris obeundum testimonio, tum illustrissimorum quorundam virorum, tum insignium medicorum, praesertim Hieronymi Mercurialis nostra hac tempestate medici absolutissimi, quique praeclarissima cum laude primas in Patavino gymnasio partes merito obtinet."

55. Augenio, *Epistolae* (1602), vol. 2, bk. 3, [19] (to Francesco Cirocco, dated from Tolentino, 1576), 44r–45v. At 44r–v: "Noster Archangelus Mercenarius nunc

ordinariam Patavii Philosophiam profitetur. Habetque collegam Franciscum Piccolomineum omnium philosophorum nostri temporis facile principem. Promotionis huius te mecum plurimum gaudere non dubito, cum te illum prosequi non vulgari amore sciam et certe dignus est amari. Ego certe non modo illum amo, sed observo plurimum tanquam decus et ornamentum omnium literatorum nostrae provinciae." On the career and writings of Mercenario (d. 1585), see Charles H. Lohr, "Renaissance Latin Aristotle Commentaries: Authors L–M," *Renaissance Quarterly* 31 (1978): 584–87.

56. Augenio, *Epistolae* (1579), 2.2, pp. 78–79 (= *Epistolae* [1602], vol. 1, 24v–25r): "Excellentissimo Horatio Augenio, Arcangelus Mercenarius . . . In qua a Mercenario proponuntur Augenio quaedam problemata philosophica, et medica discutienda"; Augenio, *Epistolae* (1579), 2.3, pp. 80–85 (= *Epistolae* [1602], vol. 1, 25r–26v): "Horatius Augenius Arcangelo Mercenario philosopho egregio . . . respondetur sigillatim ad problemata superioribus literis Augenio proposita."

57. Augenio, *Epistolae* (1602), vol. 2, bk. 5, [26] (treatise in 23 chapters, dated from Turin, 1583), 70r–85r: "Apologeticus adversus Archangelum Mercenarium Philosophiae ordinariae in Almo Gymnasio Patavino Professorem." Colombo's letter is mentioned at 70v. Mercenario's critique of Augenio's initial response is Arcangelo Mercenario, *Dilucidationum Archangeli Mercenarii, . . . volumen secundum, quo, una cum priore, universa fere naturalis scientia percurritur, necnon Horatii Augenii responsa quibusdam problematibus ab Archangelo illi propositis expenduntur deque individuationis principio Scoti defensio inanis prorsus ostenditur* (Venice: P. Meiettus, 1582), which I have not seen.

58. Orazio Augenio, *De ratione curandi per sanguinis missionem, libri decem, in quibus . . . omnia ad hoc argumentum pertinentia, secundum Galeni doctrinam explanantur . . . Addidimus eiusdem Disputationem adversus A. Mercenarium* (Turin: apud Jo. Baptistam Ratterium, 1584), 429–73 (I have not seen this edition, and cite the page nos. from Simili, "Orazio Augenio da Monte Santo"). The work against Mercenario is also included in Augenio, *Epistolae* (1602), vol. 2, bk. 5, [26], 70r–85r.

59. The explanation of the circumstances that caused him to delay his reply is found in Augenio, *Epistolae* (1602), vol. 2, bk. 5, [26], 70r–v. If this explanation is to be believed, the appearance of the year 1573 as that of the elder Augenio's death in another of Augenio's letters (*Epistolae* [1602], vol. 2, bk. 6, [33] [to Giovanni Francesco Sanzio, dated from Turin, 1581], at 98v), must be a misprint.

60. Augenio, *Epistolae* (1602), vol. 1, bk. 2, 3, 25r: "Qui fieri potest, Mercenari, ut in animum induxeris tuum, Augenium assecuturum? Utpote qui doctrina, et iudicio minime sit cum illis conferendus, homo vix inter proprios lares cognitus . . . potissimum vero qui a 27 annis continuis medicinae factitandae incumbens ab eiuscemodi contemplationem genere maxime videatur alienus?" Augenio, *Epistolae* (1602), vol. 2, bk. 5, [25], 70r: "tu vir clarissime, quo cum nunquam antea fueram locutus, ad me dedisti litteras, quibus ad disputationes quasdam

medicas, et philosophicas tecum agitandas invitare voluisti. Vix dici potest, quam fuerim admiratus. Importunum certe illud fuit scribendi genus a te, nulla prorsus accedente occasione tentatum: praesertim vero cum scires, me longe quidem ab eiuscemodi disputationibus abesse, tum quia illas longissimo intermiserim tempore, tum quia aliud erat tunc mihi negotium, quam incumbere litteris: dum vero diu, multumque cogitassem, nec invenire caussam potuissem, propter quam mihi infinitis propemodum laboribus, et animi perturbationibus, occupato, et iam discessuro, dubitationes illas proposuisses: id certe ad vires ingenii mei tentandas, effecisse te suspicabar: quod postea tuae illae, quas ad me tunc dedisti, litterae confirmarunt."

61. Augenio, *Epistolae* (1602), vol. 2, bk. 10, [61] (dated from Turin, 1583), 137r–141v: "Arcangelo Mercenario concivi suo philosophiam ordinariam in Gymnasio Patavino profitenti, viro praestantissimo. Explanatur definitio concoctionis ab Aristotle lib. quarto Meteorologicorum proposita." Aristotle's main statement of the concept is in his *Meteorologica* 4, 379b10–381b22; see also G. E. R. Lloyd, *Aristotelian Explorations* (Cambridge: Cambridge University Press, 1996), 83–103. Augenio addressed two letters on natural philosophical issues to Pietro Simone Fausto, a physician and former pupil of Mercenario: Augenio, *Epistolae* (1602), vol. 2, bk. 2, [11] (dated from Turin, 1590), 30r–31r: "Cui nam prima sphera innitatur, et an sit in loco"; and bk. 2, [12] (dated from Turin, 1590), 31r–32r: "An cutis, vel caro, vel nervus vel vor [cor], sit sensus tactus organum secundum Aristotelis sententiam expenditur." At 31r: "Maxime admiror, quod scribis: nam Archangeli Mercenarii concivis mei citra omnem controversiam philosophi praestantissimi, et de philosophia optime meriti, tanta fuit doctrina, taleque acumen ingenii, ut nihil in Aristotelis aliorumque Peripatericorum libris adeo obscurum existere censeam, ut vel ille non dilucidaverit hactenus, ut eius monumenta testantur, vel certe explicare, et (quod dici solet) at vitam resecare non potuerit: sed esto hoc problema determinare, ut dicis, non potuisse, an fortasse poterit Augenius? Qui totus Galeno, et medicinae addictus, philosophia studia iam diu intermisit." On Fausto, see Marini, *Degli archiatri pontifici*, 1:486–87.

62. Augenio, *Epistolae* (1600), vol. 3, bk. 2, 5 (to Alessandro Massaria), pp. 34–62: "De medicis novatoribus disputatio" (in the "Index epistolarum" at the beginning of the volume, this item—a treatise in seventeen chapters dated from Padua, 1598—is more pugnaciously titled "Alexandro Massariae medicinae practicae in Gymnasio Patavino doctori primario qua de novatoribus medicis agitur, omnibus calumniis, et iniuriis, et mendaciis illius hominis respondetur"). At p. 58, responding to Massaria's claims to have helped Augenio in various ways: "Magna quidem sunt haec, Massaria, turpis certe nota. Ad rem. Officia ista recenseamus. Primum officium. Quod fuerim accitus in hanc scholam Patavinam tua in primis opera atque honorifica testificatione. Falsum et temere dictum officium. Non tua opera, ut gloriaris, conduxit me in hanc scholam Senatus Venatus, sed quia meam operam iudicavit utilem, ac necessariam. Etenim scivit, quid ego in antiquissima, et augusta

Taurinensium academia, Francisci Valleriolae successor, ac primarius Practicae ordinariae doctor, utilitatis praestarem: quanta cum dignitate . . . et quanta auditorum frequentia, et utilitate meo munere perfungerer. Scivit haec Senatus non ex te, ipso harum rerum ignaro, sed ex suis oratoribus, qui me de suggesto profitentem, non raro audierunt, viris gravissimis, illustrissimis, senatoribus amplissimis . . . Alii etiam plures suam praestiterunt operam . . . In summa, Alexander Massaria, non tu me in hanc scholam conduxisti, sed labores mei, et lucubrationes meae."

63. See Augenio, *De ratione curandi per sanguinis missionem libri decem*; idem, *Disputationum de ratione curandi per sanguinis missionem ex Galeni sententia libri septem;* idem, *De ratione curandi per sanguinis missionem libri XVII in duos tomos divisi.*

64. Augenio, *Epistolae* (1600), vol. 3, bk. 12, 45, p. 475: "Agitur de obitu Massariae [d. 1598] . . . Fuit mihi acerbissima eiuscemodi mors, tum quia tenebat me ille in medicis studiis continuo exercitatum, tum quia desiderabam, ut responsionem legeret ad suum additamentum apologeticum, quam parabam."

65. Girolamo Mercuriale, *Responsorum et consultationum medicinalium tomus quartus. Nunc primum a Guilielmo Anthenio philosopho et medico editus* (Venice: apud Iuntas, 1604; subsequently reprinted in idem, *Consultationes et responsa medicinalia quattuor tomis comprehensa* [Venice: apud Iuntas, 1624]), consilium 10, pp. 21–24: "Hieronymo Fabricio. Hieronymus Mercurialis S.D. De commentariis Horatii Augenii de sanguinis missione." At p. 21: "Cum hisce proximis diebus commentationes Horatii Augenii viri doctissimi, atque mei amicissimi, de sanguinis missione perlegerem, ac in eam disputationem de iudicationibus [*sic:* indicationibus] mittendi sanguinis a compluribus diserte satis tractatam, nec dum finitam incidissem, magnopere sum miratus, ut fiat, quod saepenumero ingenia praeclara, cum rebus alioquin per se conspicuis maiorem afferre lucem contendunt, eas obscuriores, et difficiliores rendant, quod eo magis viri eximii atque spectatae authoritatis omni studio evitare debent, quo ipsi cum multos sequaces, sibique omnia credentes habeant, eo pluribus non sine aliquo artis dedecore errandi, vel saltem semper instar Pyrrhonicorum haesitandi occasionem praebent, praeterquam quod ab illo aeterno Aristotelis in libro *Rhetorica* praecepto abscedunt, ubi docens homines naturaliter optare facilitatem, et brevitatem dicendi . . . simul iubet, ne qui docere alios studet, longus, difficilis, atque obscurus sit: et certe (libere loquar) in hac de mittendo sanguine controversia, a nonnullis tamen peritissimis brevissime examinata, vix possum ferre, ingenii, et eloquentiae tantum consumi, nec aliud quidquam tandem obtineri, quam libidinem quandam immanem pugnandi, ac non sine aliqua animorum studiosorum pernitie, disputationes protrahendi, cum praesertim paucis totam litam terminare, si contentiones abijciamus, nobis liceat."

66. On this speech, see Antonio Riccoboni, *De gymnasio Patavino commentariorum libri sex* (Padua: apud Franciscum Bolzetam, 1598), 71r, and Nancy G.

Siraisi, "Giovanni Argenterio and Sixteenth-Century Medical Innovation: Between Princely Patronage and Academic Controversy," *Osiris* 6 (1990): 161–62.

67. See Augenio, *Epistolae* (1600), vol. 3, bk. 2, 5 (to Alessandro Massaria), pp. 34–62: "De medicis novatoribus disputatio."

68. Augenio, *Epistolae* (1579), 3.1 (to Giovanni Battista Cucina), p. 86: "Ioannis Argenterii (quem semper tum ob insignem eius doctrinam, et liberam docendi professionem, tum ob amorem erga me ingentem, honoris causa nominabo)"; Augenio, *Epistolae* (1602), vol. 2, bk. 3, [19] (to Francesco Cirocco, dated from Tolentino, 1576), 44r: "Non sum editurus commentarios meos in Galeni librum ad Glauconem, et de febrium differentiis, licet manum extremam habeant, quod ad me pertinet: nisi prius novero, quae de hoc argumento scripserit Argenterius. Nam puto illum omnem movisse punctum, nullumque mihi reliquisse locum." Argenterio had died in 1572; his work on fevers was first published (posthumously) in Giovanni Argenterio, *Opera nunquam excusa, jamdiu desiderata, ac e tenebris in lucem prodita. In duas partes distincta. Quarum prior commentarios in Hippocratis Aphorismorum primam, secundam, & quartam sectiones plus sex, & triginta annorum spatio elaboratos: altera vero De febribus tractatum singularem: & primi libri [Galeni] ad Glauconem praeclaras explanationes: item De calidi significationibus, ac calido nativo doctissimum libellum complectitur* (Venice: apud Juntas, 1606). Augenio's *De febribus . . . libri septem* was first published (also posthumously) in 1604.

69. Augenio, *Epistolae* (1600), vol. 3, bk. 2, 5 (to Alessandro Massaria), pp. 34–62: "De medicis novatoribus disputatio." In cap. 2, pp. 35–37: "Alexandrum Massarium nunquam fuisse ab Augenio inter novatores medicos repositum," at p. 37: "Cur igitur, inquies, mecum disputas? Cur meis rationibus satisfacere conaris? Quia, inquam, connivere, et assentiri, ubi de hominis agitur vita, ut saepe dixisse memini, est impium, atque detestabile. Quis aequo animo ferre possit cum Hippocratis, antiquorumque omnium instituta subverti videat? Cum Galeni doctrinam, cuius te ad unguem imitatorem esse profiteris, aperte corrumpas? Cum comunem omnium Arabum, omnium Latinorum opinionem contemnis, ac detesteris? Cum omnes, qui de indicationibus hucusque scripserunt, turpiter fuisse deceptos, et somno habuisse conniventes oculos profitearis? Cum tibi persuadeas licere passim mutilare Galeni verba? Imo sententias integras? Imo falsos citare contextus? Cum de omnibus huius almi gymnasii professoribus tam male sentias? Cum ausus sis publice profiteri literatorum hominum eos minutulos esse simios: cum tu solus velis in hac schola reputari alter Galenus? Cum mihi passim, atque aliis imponas apertissime?"

70. On this proposal (made in 1600) and its outcome, see Nancy G. Siraisi, *Avicenna in Renaissance Italy* (Princeton: Princeton University Press, 1987), 113–20. As a lecture on the relevant portion of the *Canon* had previously existed, Augenio may have thought it easier to revive a formerly existing chair than to introduce a completely new one.

71. Augenio, *Epistolae* (1600), vol. 3, bk. 2, 5 (to Alessandro Massaria), pp. 34–62: "De medicis novatoribus disputatio." In cap. 2, pp. 35–37, at p. 35: "ostendam qui ex mea sententia sint novatores, quodque hoc nomen odiosum et detestandum non sit." The list of innovators on p. 36 includes Pierre Brissot, Laurent Joubert, and others, "viros quidem omni scientiarum genere ornatissimos, modestissimos, prudentissimosque." In cap. 8 ("Novatores non esse corruptores reipublicae literariae"), pp. 44–45, at 45: "Nam si novator est, qui volens a Galenis discedit doctrina, omnisque novator reipublicae literariae corruptor debet nuncupari, et homo pestilens, omnino sequitur viros praestantissimos, quos semper omnia venerata est aetas, vitiosissimos ex tua sententia extitisse, qui contra Galenum disputarunt: cuiusmodi sunt Averroes, Avicenna, Gregorius Naziazenus, D. Thomas, et universa theologorum schola in iis, qua ad animam pertinent, a Galeno dissentiens, Alphonsus a castro, Ioannes Picus Mirandulae, et Concordiae Comes, Hieronymus Francastorius, at alii quamplures, quos brevitatis causa non recenseo. At maxima mihi quidem certe videtur blasphemia pestilentes hosce appellare viros, ac reipublicae literariae corruptores."

72. For example, Augenio, *Epistolae* (1600), vol. 3, bk. 2, 5 (to Alessandro Massaria), "De medicis novatoribus disputatio," cap. 5 ("De Argenterio, et Fernelio"), pp. 40–42: "Quae cum ita habeant, cur me arguis, quod dixerim . . . maximas habendas gratias Argenterio, quod contra Galenum docte et subtiliter scripserit? Nam aut vera sunt, quae scripsit, aut falsa. Si vera, cur non habebimus gratias, quod delitescentem veritatem invenerit? Si falsa, cum sint acute, et probabiliter dicata, gratiae debentur, quod aliorum ad inveniendum veritatem excitaverit ingenia."

73. Augenio, *Epistolae* (1600), vol. 3, bk. 2, 5, cap. 4, pp. 37–39: "Qui sint in universum novatores appellandi, qui novatores medici, et aliarum scientiarum professores. Nomenque novatoris non esse turpe ac detestandum semper et usquequaque." At pp. 38–39: "Praeterea novatores medici appellari consueverunt, qui a Galeni doctrina aliqua ex parte volentes, discesserunt. Diximus, aliqua ex parte, ob eos qui non dissentiunt in tota arte, sed in quadam particula, circa quam novam afferre conantur doctrinam. Inter antiquos medicos novatores fortassis primi fuerunt, Avicenna et Averroes. Ille quidem rarius, hic vero frequentius Galenum arguit.

"Inter medicos recentiores, ut praetermittam multos, duo fuerunt antesignani: Ioannes Argenterius, et Ioannes Fernelius. Hic modeste et doctissime, et subtiliter, adversus Galenum aliquando disputat: ille autem acerbius, et acrius multo atque frequentius a Galeno dissentit. At vero in fundamentis, ac in eisdem principiis conveniunt, nimirum quod omnes dogmaticam, et rationalem profiteantur medicinam.

"Postremo novatores vocantur artifices, qui plerisque omnibus veterum placitis reprobatis, novam artem, atque scientiam introduxerit. In quorum numero primus novator fuit Hippocrates, secundus Aristoteles, tertius Philippus Theophrastus Paracelsus eremita. Hippocrates libro de hominis natura veterum opinionibus de principiis rerum naturalium reiectis principia posuit medicinae rationalis: Aristotelis autem antiquorum plurimis reprobatis opinionibus inventor fuit logicae philophiae, naturalis et divinae, ut Averroes ait in praefatione libri physicorum. Utrunque Galenus imitatus est, qui acerrime contra veteres ac eos, qui sua aetate vivebant, erantque in summo precio, ut medicam artem perficeret, perpetuo disseruit. Eremita ille novam condidit sectam, medicinam scilicet chymicam, ut vocant. Quam plures enim ex eius discipulis contendunt ipsum artem invenisse destillatoriam. Ego autem licet facile admittam illius medicinam vere novam esse, a Galeni institutis alienam, non tamen, ipsum censeo destillationis artem invenisse. Chymica enim ars antea inventa fuit, quemadmodum quamplures animadverterunt, praesertim Andernacus libro de veteri et nova medicina. Zosimus enim, Blamidas, Olympiodorus, Eben Mesue, et alii praeclare ostendunt chymicam exteros populos in usu habuisse, cui suffragari videtur Galenus libro de Theriaca ad Pisonem, ubi ipse medicamentorum per ignem paratorum usum ostendit, commendatque. Sed tamen hoc laudis Paracelso merito tribuitur, quod eandem artem hoc nostro seculo primus fere in medicum usum revocasse videatur, et non mediocriter ampliasse. Siquidem praeclare docet, quomodo hac ipsa arte aquae, liquores, olea, sales, non solum ex herbis, lignis, fructibus, sed etiam ex succis concretis, terrae fossilibus, unionibus, corallis, lapidibus pretiosis, aliisque id genus solidis corporibus extrahantur, nec non sublimes quoque spiritus, facultates, et quintae quasi essentiae ex eisdem deducantur, atque in pessimorum morborum usum recte possint accommodari, imo vero pro totius corporis praeservatione etiam utiliter administrantur. Et propter hanc caussam, ut olim qui aliis non cedunt remediis, ante illos autem totius corporis curandi gratia exhibentur, artem hanc doctissimi, gravissimique viri exercebant: sic nostra hac aetate magni et sublimes heroes exercent et colunt. Tot igitur modis novator quispiam dicitur artifex."

74. Augenio, *Epistolae* (1602), vol. 2, bk. 4, [9], 60v; Nancy G. Siraisi, "Hermes among the Physicians," in *Das Ende des Hermetismus: historische Kritik und neue Naturphilosophie in der Spätrenaissance*, ed. Martin Mulsow (Tübingen: Mohr Siebeck, 2002), 209; Richard Palmer, "Pharmacy in the Republic of Venice," in *The Medical Renaissance of the Sixteenth Century*, ed. A. Wear et al. (Cambridge: Cambridge University Press, 1985), 111, 116.

75. For example, see Augenio, *Epistolae* (1600), vol. 3, *3r–v; idem, *Epistolae* (1602), vol. 1, bk. 8, 1, 86r.

76. Augenio, *Epistolae* (1600), vol. 3, bk. 2, 5 (to Alessandro Massaria), pp. 34–62: "De medicis novatoribus disputatio," in cap. 2, p. 35.

77. Ian Maclean, *Learning and the Market Place: Essays in the History of the Early Modern Book* (Leiden: Brill, 2009), "André Wechel at Frankfurt, 1572–1581," 163–225.

Conclusion

1. Giovanni Manardo, *En postremum tibi damus, candide lector . . . epistolarum medicinalium libros XX. e quibus ultimi duo in hac editione primum accesserunt, una cum epistola iandudum desiderata, de morbis interioribus, quam utinam immatura morte non praeventus, totam absolvere potuisset. Eiusdem in Ioan. Mesue Simplicia & composita annotationes & censurae, omnibus practicae studiosis adeo necessariae, ut sine harum cognitione aegrotantibus recte consulere nemo possit* (Basel: Apud Mich. Isingrinium, 1540), letter of dedication (dated at end 1529), α4r–v: "Reverendo ac magnifico Alphonsino Trotto, equiti Hierosolymitano, Coelius Calcagninus." At α4v: "Certe oratio ita perspicua est, ut neminem torqueat: ita pura, ut neminem, nisi paulo delicatiorem, offendat. Nam quod ex consuetudine medicorum, nonnulla in eorum libris protrita interserverit, quae aliquis fortasse morosus aversetur: par est ut, quisquis ille est, secum cogitet, non modo a quo, sed ad quos etiam scribatur: et orationem non in dicentis, sed in audientis multo maxime gratiam repertam esse . . . Nec sane quispiam putet Galenum, Oribasium, Aeginetam, a Manardo accuratius excussos, quam Ciceronem, Demosthenem, Plinium, caeterosque elegantioris notae."

INDEX

Page numbers in italics refer to figures.

Aeschylus, 45
Aetius, 47
Agricola, Georg, 31
Agrippa, Heinrich Cornelius, 57
Albertus Magnus, 32
alchemy: and Gesner, 5; and Lange, 44, 46, 52; and Monau, 32; and natural philosophy, 2; and Ottheinrich, 44, 58. *See also* Paracelsianism
anatomy, 17, 70–71, 75, 87
Ancina, Juvenal, 72
antiquarianism and antiquities: and Augenio, 69; and humanism, 8–9, 85; Lange's interest in, 42–43, 46, 55, 60
anti-Semitism: and accusations of magic, 56–58; and ritual murder accusations and trials, 56–68
Antoninus Pius, 50–51
Argenterio, Giovanni, 22, 23, 80–81, 82, 134n4
Aristotelianism, 2; and Augenio, 77–78, 78–79, 82; and Lange, 40, 48, 56; and Massa, 18; and Mercuriale, 31, 32, 80; and Monau, 32; and Gianfrancesco Pico, 58; at University of Padua, 19, 77–78
astrology, 43–44, 45, 60, 73, 143n40
Athenaeus, 45
Auerbach, Heinrich Stromer von, 18–19, 52

Augenio, Fabrizio, 70
Augenio, Lodovico, 70
Augenio, Orazio, 62, 63–84; and Argenterio, 80–81, 82, 134n4; and astrology, 73, 143n40; and Avicenna, 81, 82, 151n70; biography of, 63–66, 134n5; bloodletting treatise of, 64, 68, 78, 83, 136n8; commitment to publishing letters of, 9–10, 64–65, 83–84; as consultant and expert witness to ecclesiastical and secular officials and courts, 72–76, 83; correspondence of, 66–70, 72, 74, 139n21; *Epistolae medicinales*, 10, 64, 66, 75; female anatomy views of, 71, 75–76; and Fernel, 68; fevers treatise of, 64, 65, 68; and innovation, 63, 68, 71, 76, 80–83, 87; interest in wine of, 136n9; on kidney and bladder conditions, 68, 71; and Massaria, 79–81; medical innovation treatise of, 81–83; and Mercenario, 77–79; and Mercuriale, 77, 79; Padua connections of, 64–65, 66–67, 76–83, 87; pregnancy and childbirth treatise of, 64, 68; testimony in matrimonial dispute by, 74–75. *See also* controversies
Augsburg, Diet of (1548), 51, 125n47
Averroës, 43, 65, 82

Avicenna, *Canon:* and Augenio, 81, 82, 151n70; disease theory in, 23; and Lange, 43; and Monau, 32; as standard medical text, 17, 24, 43, 81, 123n42

Bacci, Andrea, 70, 141n34
Barbaro, Ermolao, 96n18
Basel, publishers in, 6, 21, 28, *29,* 42, 53
Basel University: and interregional correspondence, 16; and Monau, 23, 34; Zwinger's correspondence and, 3–4, 27, 28, *29,* 74
Bauhin, Gaspard, 24–25
Bavaria, 26, 34
Bech, Philip, 53, 125n52
bloodletting: Augenio's treatise on, 64, 68, 78, 83, 136n8; controversies surrounding, 71, 79–80
Bologna University, 41, 55, 65
Bonaventura, Giacomo, 27
botany, medicinal, 46–47, 69, 73–74, 140n29
Bozzuto, Annibale (Cardinal), 72–73, 143n40
Brescia, 5–6
Burchard, Franz, 96n18

cabala, 57, 58
Calcagnini, Celio, 85
Calvinism, 50
Camerino, 64
Capodivacca, Girolamo: and Augenio, 76–77, 79; Erastus correspondence of, 21–23, 24, 35
Carben, Victor von, 131n80
Cardano, Girolamo, 20, 121n32
Carlo Emanuele, Duke of Savoy, 65
Carlstadt, Andreas, 41, 53
case histories and *observationes,* medical, 7, 8, 48, 60, 75
Castiglioni, Angelo, 6
Catholic Church: Augenio and ecclesiastical authorities of, 72–73, 74–76; and censorship, 15–16; enforcement of religious conformity

by, 18–19, 19–20, 35; Lange's hostility to, 53–54; vs. Paracelsianism, 34
Celsus, Aulus Cornelius, 27
censorship, 15–16, 28, 35, 86
Charles II, Archduke of Austria, 26, 107n36
Charles V, Emperor, 39, 48
Christian Hebraism, 57, 59
Cibo, Vicenzo, 69
Cicero, Marcus Tullius, 7, 8, 40
Cingoli, 67, 139n21
Cini, Giulio, 68–69, 71, 77
Cirocco, Francesco, 67
Clement VII (Pope), 70
Clement VIII (Pope), 71
Clusius, Carolus (Charles de l'Écluse), 4
Codronchi, Giovanni Battista, 27
collecting and collections, 6–7, 8–9, 11, 28, 46, 83
collections of medical letters (as genre), 8–12; anthologies by multiple authors, 9, 11; Augenio's attachment to this form of publication, 64–66; contemporary descriptions of, 39; and humanism, 8–9, 85; informal manuscript circulation of individual letters, 1, 10–11, 15–16, 67; initiation of printed genre, 1; purposes/value of, 3–6, 11–12, 85–86; as single-author works, 9–10
Colombo, Michele, 25, 75, 78
Colombo, Realdo, 71
complexio, theory of, 22–23, 31
consilia: by Augenio, 65–66, 74; by Da Monte, 4; functions of, 6, 7; by Lange, 48; by Massa, 18, 20; and medical letters, 8, 25, 86; by Mercuriale, 25–27
contagion, discussions of, 21–23, 30–31, 73
controversies: Augenio's involvement in, 66, 83, 87; with Cini, 68–69, 71, 77; with Massaria, 65, 79–82
controversies, medical. *See* bloodletting; contagion; debate and disputation; innovation, medical

Cordella, Girolamo, 71, 140n28, 142n35
correspondence networks: and court
physicians, 26, 50–51; functions of,
4; geographic range of Venetian
territories and, 19; motivations for
participating in, 16–17; and professional advancement, 66
cosmography, 41
courts, princely, medicine and physicians
at: and career progression, 10; and
correspondence networks, 26, 50–51;
Crato as court physician, 26; influence
of physicians at, 40, 46, 47–50, 86,
120nn31–32; interregional correspondence of, 16, 30. *See also* Lange,
Johann
Crato von Krafftheim, Johann:
anti-Paracelsianism of, 34–35;
correspondence publication of (by
Scholz), 10, 20, 27–28, 35; and Gesner,
4–5; interregional correspondence of,
20–21, 26, 53; and Mercuriale, 26, 27,
30–31, 107n35; reputation/renown of, 7
Crispo, Pietro, 66, 71, 138n18

d'Abano, Pietro, 43
Da Monte, Gian Battista, 4, 134n4
debate and disputation: in Monau's
letters, 32–34; value of letters in
fostering, 8, 9, 16, 66, 83, 86. *See also*
controversies
Delisle, Candice, 1
demons and demonic possession, 57; and
ghosts, 54, 126n60; Lange's views on,
46, 54–55, 128n64; and Paracelsus, 5.
See also magic (natural and demonic)
diet/food, 45, 49, 53, 125n47
Diodorus Siculus, 45
diseases, illnesses, injuries: chlorosis/
green sickness, 121n34; dysentery, 30;
fevers, 48, 64, 65, 68; hemorrhage, 74;
impotence, 75; kidney stones, 30, 69,
71; plague, 64; pleurisy, 71; poisons,
24, 33, 45, 49, 73; syphilis *(morbus
gallicus)*, 17, 18, 24; urinary conditions,

30, 45, 48, 68, 69, 71–72, 76–77;
wounds, 48
disease transmission, theories of, 21–23,
30–31, 73
Dodoens, Rembert, 30, 34
Donzellini, Girolamo, 35–36
drinking of alcoholic beverages, 45, 49
Dudith, Andreas, 30
Duno, Taddeo, 19–20

eating. *See* diet/food
Eberhard, count of Erbach, 126n58
Eck, Johann, 41, 53
Efron, John M., 132n82
Egypt (ancient): alchemical and other
medicaments attributed to, 44, 45;
natural magic in, 46, 54, 57, 58
Emanuele Filiberto, Duke of Savoy, 64,
74
empiric medical practitioners, 63, 70, 83
epidemics, 21, 23, 30–31, 48, 73
*Epistolae medicinales diversorum
authorum* (1556), 9, 11, *14*, 42
Erasmus, Desiderius, 5, 9
Erastus, Thomas: anti-Paracelsianism of,
21, 34–35; and correspondence with
Padua professors, 21–23, 28, 34–35;
*Disputationum et epistolarum medicina-
lium volumen*, 7, 12, 21; and Heidelberg University reform, 50
Eustachi, Bartolomeo, 70, 71, 75, 141n34

Fabrizi d'Acquapendente, Girolamo, 27,
79–80
Farnese, Alessandro (Cardinal), 24, 71
Ferdinand I, Grand Duke of Tuscany, 24
Ferdinand of Austria, 51
Fernel, Jean, 22–23, 68, 80–81, 82
Ficino, Marsilio, 57
Finetti, Giustiniano, 134n4
Fontana, Lavinia, 105n27
Forster, Georg, 53, 125n51
Fracastoro, Girolamo, 31
Franceschini, Egidio, 69, 140nn28–29
Frankfurt, publishing in, 42

Friedrich II, Elector Palatine: and Jewish
settlement / medical practice, 59,
133n85; Lange's service to, 41, 48, 50,
51, 124n47; Lutheranism of, 49–50,
53–54; medical recipe collection of,
124n45
Friedrich III, Elector Palatine, 50

Gaetani, Enrico (Cardinal), 26
Galenism: and Aristotelianism, 77–78,
78–79; and Augenio, 65, 69, 77–78,
81; and bloodletting, 80; departures
from, 2; disease theory of, 22–23; and
humanism, 80; and innovation, 81–82;
and Lange, 42–43, 44, 47, 50, 55, 56;
matter theory of, 31–32; vs. Paracelsianism, 34–36, 82–83; and purgatives, 68. *See also* Hippocrates
Galileo Galilei, 25
Gasser, Achilles Pirmin, 53
Geraldus (student of Lange), 45, 56
German Nation (*Natio germanica
artistarum,* association of northern
European students at Padua University), 19, 21, 24–25
Germany and German lands (linguistic
region; imperial, territorial, and
city polities): Heidelberg (city and
Palatinate court at), 47–51, 59; Jewish
settlement in, 59–60, 132–33nn84–86;
medical education in, 18; printing/
publication in, 2, 11–12, 20, 36, 86. *See
also* Heidelberg University; Leipzig
University; Palatinate
Gesner, Conrad, 29; and alchymical
recipes, 92n11; *Bibliotheca universalis,*
29; Crato correspondence of, 4–5;
Epistolae medicinales, 4; Lange
correspondence of, 53; Mondella
correspondence of, 5–6; and natural
history, 31; and publication of his own
letters, 1, 10, 12; and publication of
Lange's letters on surgery, 42; as
significant writer of published letters,
3, 4, 7, 83

Gilly, Carlos, 34
Giraldi, Lilio Gregorio, 29

Heidelberg (city and Palatinate court at),
47–51, 59. *See also* Lange, Johann
Heidelberg University, 7, 16, 21, 118n22;
and Lange, 50–51; organizational
and curricular reform of, 49–50, 60;
reformed religion in, 34
Herodotus, 45
Hippocrates: Augenio on, 81, 82;
Lange on, 42, 43, 44, 50, 56; letters
attributed to, 7; Mercuriale's edition
of Hippocratic corpus, 24, 33; and
Paracelsianism, 34; remedies of, 69
Homer, 47
humanism: and enthusiasm for letter
collections, 8–9, 85; and Galenism, 80;
Lange's, 40, 42–47, 60, 63, 86–87
human life span, 52

innovation, medical: and Argenterio,
22–23; and Augenio, 66, 68, 71, 76,
80–83, 87; and Fernel, 22–23, 68,
80–81, 82
interregional correspondence, 2, 7, 86; of
Augenio, 72, 74; and court physicians,
16, 30; of Crato, 20–21, 26, 53; and
religious differences, 18–20, 28,
34–46
Ipsilla, Pietro of Egina, 41
Italy (geographic region, states, and
cities): censorship in, 35; epidemics in,
21; Le Marche, 63, 66, 67, 70, 72, 83;
medical education in, 10, 41; Monte
Santo, 63, 77, 78; Papal States, 63, 70,
83; printing/publication in, 20, 26–27;
religious atmosphere in, 5–6; Rome,
74–75, 75–76; as source of printed
collections of medical letters, 2.
See also Padua University; Venice

Jewish medical practitioners: in
Germany, 133n85; in Italy, 123n43; in
Palatinate, 59; restrictions on and

hostility toward, 50, 56–57, 58–60, 132n82
Jewish settlement in German states and cities, 59–60, 132–33nn84–86
Joubert, Laurent, 76

Kolreuter, Sigismund, 72
Kress, Christoph, 48, 51, 125n47

Lange, Johann, *38,* 39–60; antiquarianism of, 42–43, 46, 55, 60; and Augenio, 65; biography of, 40–42; Bologna studies of, 41; vs. Catholic beliefs and practices, 53–54; on chlorosis / green sickness, 121n34; contacts and correspondents of, 50–53; as court physician, 46, 47–50, 60; at Diet of Speyer, 124n47; *Epistolarum medicinalium volumen tripartitum* (1589), 39; *Epistolarum medicinalium volumen tripartitum* (1605), 38, 42; family network of, 39–40; vs. Jewish medical practitioners, 50, 56–57, 58–60; Leipzig University association and, 40–41, 50, 51, 52, 117n14, 125n52; Lutheran adherence of, 53–54, 124n46; and magic (natural and demonic), 46, 54–59, 128n64; medical humanism of, 40, 42–47, 60, 63, 86–87; vs. medical practitioners without university training, 56; *Medicinalium epistolarum miscellanea* (1554), 42; *Medicum de republica symposium,* 42, 50, 51, 53–54, 56, 126n58; and natural history, 46–47; oration at Leipzig disputation by, 41; and Gianfrancesco Pico, 41, 53, 57–58, 130n73; recipes of, 49, 122n38; *Secunda epistolarum medicinalium miscellanea* (1560), 38, 42; as significant writer of medical letters, 7, 11–12, 39, 42, 86–87; travel diary of, 41
Lazius, Wolfgang, 120n32
Leiden University, 4
Leipzig University: disputation at, 41; Lange's studies, teaching, and

colleagues at, 40–41, 50, 51, 52, 117n14, 125n52; Lutheranism and, 18
Le Marche, 63, 66, 67, 70, 72, 83
Leonibus, Ludovicus de, 41
Leoniceno, Niccolò, 41
Leo X, Pope (Giovanni de' Medici), 41
letter writing, ancient models of: and brevity, 27; examples of, 4–5; humanist enthusiasm for, 8–9, 60, 85; Scholz's regard for, 6–7
Lipsius, Justus, 9, 46
Lonie, Iain, 68
Lotichius Secundus, Petrus, 51
Louis II, King of Hungary and Bohemia, 40
Love, Harold, 99n33
love philtres, 30
Ludwig V (Count Palatine of the Rhine), 59, 124n45, 133n85
Luther, Martin, 41, 53, 59
Lutheranism, 18–19, 49–50, 53–54, 87, 124n46
Lyon, publishing in, 42

Macerata, University of, 63–64
Maclean, Ian, 1, 8, 12, 84, 86
Mader, Theophilus, 12, 21
Mader, Timotheus, 21
magic (natural and demonic): attributed to biblical figures, 54–55, 57, 58, 126n61; attributed to Jews, 57; as cause of impotence, 75; Lange on, 46, 54–59; Mercuriale on, 27, 30; Gianfrancesco Pico on, 57–58, 130n75; supposed historical corruption of, 54, 57–58, 126n60, 129n68, 132n81. *See also* demons and demonic possession
Manardo, Giovanni: and Augenio, 65; humanism of, 85; on letters vs. *consilia,* 8; as significant writer of medical letters, 1, 3, 39, 43, 60
Margaritha, Antonius, 59, 131n80
Massa, Apollonio, 17–19, 36
Massa, Lorenzo, 17–18

Massa, Nicolò, 17–20, 39, 52; and
anatomy, 17; Auerbach correspondence of, 18–19; and Avicenna, 17–18;
consilia, 17–18; *Epistolae medicinales et
philosophicae* (1550, 1558), 17
Massaria, Alessandro, 65, 79–82
matrimonial dispute, Augenio's
testimony in, 75
Mattioli, Pietro Andrea: Dioscorides
commentary of, 31; and Gesner, 4;
published collection of medical letters
of, 9–10, 12, 39, 83; views on compo-
nents of lead, 31–32
Maximilian II, Holy Roman Emperor,
24, 26, 27, 30, 34
medical practitioners without university
training, 44, 49, 50, 56; empirics, 63,
70, 83
medicaments, 5, 45, 82. *See also* recipes,
medicinal
Melanchthon, Philip, 6–7, 34, 53
Mercenario, Arcangelo, 70, 77–79
Mercuriale, Girolamo, 24–34; *De arte
gymnastica libri sex,* 24; and Augenio,
77, 79–80; *Consultationes et responsa,*
25–27, 28, 86, 106n33; and Crato, 26,
27, 30–31, 107n35; and edition of
Hippocratic corpus, 25; and lectures
on diseases of women and the skin,
poisons, and plague, 25; Monau
correspondence of, 27–28, 31–34; and
Paracelsianism, 34–35; portrait of, 24;
Zwinger correspondence of, 17, 27–28,
29, 33, 34, 36, 74
Mermann, Thomas, 26, 34
midwives, criticism of, 76
mineralogy, 31–33
Moffett, Thomas, 112n60
Moibanus, Johannes, 53, 125n51
Monau, Peter: biography of, 104n23;
Capodivacca correspondence of, 23;
interregional correspondence of,
20–21, 53; Mercuriale correspondence
of, 27–28, 31–34; and Paracelsianism,
34, 112n60

Mondella, Luigi, 5–6, 39, 55, 93n12;
Medical dialogues (1551), 5; medical
letters of, 5
Monte Santo, 63, 77, 78

Natio germanica artistarum. See German
Nation
natural history: humanist enthusiasm
for, 60; medicinal botany, 46–47, 69,
73–74, 140n29; mineralogy, 31–33
natural magic. *See* magic (natural and
demonic)
natural philosophy, 8; and Augenio, 70,
77–78, 149n61; changes in, 2; and
humanism, 43; in Mercuriale's letters,
28–31; as necessary to medicine, 56;
and religious repression, 20; and
Republic of Letters, 16, 40. *See also*
Aristotelianism
Neri, S. Filippo, 71, 72
Nifo, Agostino, 19
Nutton, Vivian, 31, 56

Ochino, Bernardino, 19
Olivieri, Serafino, 75–76, 146n50
Oporinus, Johannes, 42, 53
Ottheinrich, Elector Palatine: and
alchemy, 44, 58; as book collector, 46;
Heidelberg University reform and,
49–50; and Jewish settlement, 59–60,
133n86; Lange's service to, 46, 48,
50, 55

Padua University, 87; Augenio's
connections and career at, 64–65,
66–67, 76–83, 87; correspondence of
medical faculty with physicians in
German lands, 20–23, 28, 34–35;
Jewish students at, 123n43; and Lange,
41; northern European students at, 19,
21, 24–25; and Paracelsianism, 34–36;
renown of, 16, 19, 64
Palatinate: informally trained medical
practitioners in, 44; insect pests of, 48;
Jewish settlement in, 59; Lange's

regional correspondence network in, 51–53; Lange's service to Elector of, 40, 41, 49–50, 51; library of, 46; Lutheranism/Calvinism in, 49–50, 53, 87; medicinal plants of, 47
Palmer, Richard, 17, 35–36, 82
Papal States, 63, 70, 83
Paracelsianism, 2, 112n60; Erastus's opposition to, 21, 34–35; vs. Galenism, 34–36, 82–83; Gesner's opposition to, 5; and Jewish medicine, 58. *See also* alchemy
Paracelsus. *See* Paracelsianism
Paterno, Bernardino, 21, 22; commentary on *Canon* of Avicenna, 24
Paul III (Pope), 70
Pavia, University of, 20
Pennuto, Concetta, 144n40
peregrinatio medica, 17
Perna, Pietro, 4, 21, 28, 29, 34–35
Petrarch, 9
Petroni, Alessandro, 71, 72
Philip, Landgrave of Hesse, 133n85
Philip II of Spain, 39
Pico, Gianfrancesco, 41, 53, 57–58, 130n73, 130n75
Pico della Mirandola, Giovanni, 6–7, 46, 57, 58, 59
Piedmont, 76, 83; Senate of, 73–74
Pinelli, Gian Vicenzo, 25
Pisa, University of, 22, 41, 64, 80–81
Plato, 7, 44–45, 48
Pliny, 8, 40
Plutarch, 45, 47
Pomata, Gianna, 8
Pomponazzi, Pietro, 18, 19, 41, 55
Portaleone, Abraham, 25
Porto, Antonio, 70–71, 141n34
Prague, 16
pregnancy and childbirth, 64, 68, 72–73, 143n40
printers and publishers: and interest in medical letters, 11, 12, 14, 85–86; Junta, 11; and Mercuriale, 25–27; Oporinus, 42, 53; Perna, 4, 21, 28,

29, 34–35; Waldkirch, 28; Wechel, 11–12, 84
Prob, Christoph, 49
Proclus, pseudo- *(De sphaera),* 40–41
professional advancement and letter writing: as motive for correspondence, 16–17; reputation building and, 20, 63, 72, 76–79, 83–84, 85
Ptolemy Philadelphus, 46
purgation and purgatives: cautions about, 67, 68; and infants, 33; and plant identification, 47; powers of, 50–51

Raphanus, Wenceslaus, 30
Rascalon, Guillaume, 51, 55
Rasis, 123n42
recipes, medicinal: ancient, 69; collectors of, 49, 59, 118n22, 124n45, 133n85; different for rich and poor, 5, 55; Donzellini's, 35; and Gesner, 92n11; Lange's, 49, 122n38; physicians preparing own, 73–74
religious differences and intellectual life, 2, 6; censorship, 86; and correspondence hazards, 15–16; interregional communication, 18–20, 28, 34–36; and Lange, 40, 53–54; at Padua, 24; University of Heidelberg reform, 49–50, 60. *See also* Jewish medical practitioners
remedies, 5, 45, 82. *See also* recipes, medicinal
Republic of Letters: characteristics attributed to, 4–5, 16, 85; and free communication, 3; and innovation, 82; interregional extent of / crossing boundaries and, 7; limits/constraints on, 36; Scholz collection's position in, 7, 9; study of, 1, 90n1
reputation. *See* professional advancement and letter writing
Reuchlin, Johannes, 57, 58, 59
Reusch von Eschenbach, Johann, 52, 125n50

Reusner, Bartholomaeus, 40
Reusner, Christoph, 40
Reusner, Hieronymus, 40
Reusner, Nicholas, 39–40, 42
Richardson, Brian, 99n33
rivalry, professional, 65–66, 68–69, 71,
   77, 79–83, 87
Rome, 74–75, 75–76
Rome, University (La Sapienza), 64, 66,
   70–72
Rondelet, Guillaume, 4
Rota Romana, Augenio and, 75–76
Ruderman, David B., 132n82
Rudolf II, Holy Roman Emperor, 26, 31
Rummel, Erika, 96n18

Sanzio, Giovanni Francesco, 67
Savonarola, Michele, 120n32
Savoy, duchy of, 73–74
Scholz von Rosenau, Lorenz, 85; and
   Capodivacca's published letter, 23;
   *Consiliorum medicinalium . . . liber
   singularis* (1598), 6; Crato, *Consiliorum
   et epistolarum medicinalium liber,*
   ed. Scholz (1591–1611), 10; Crato's
   published letters, 10, 20, 27–28, 35;
   enthusiastic letter collecting of, 11;
   *Epistolarum philosophicarum,
   medicinalium, et chymicarum . . .
   volumen* (1598), 6; Mercuriale's
   published letters, 30–34; and
   publication of letters from multiple
   authors, 9, 12; on value of letters, 6–7,
   8, 9
scribal publication, 98n33
Scutellari, Giacomo, 26
Sextus Empiricus, 58
Simmelbecker, Theodoric, 21, 22
Simon of Trent, 56–57
Speyer, Diets of (1542 and 1544), 51,
   124n47
Stenglin, Lukas, 26
surgery and surgeons: Augenio on,
   71–72; Lange on, 42, 43, 44, 45, 50,
   56, 60

temperament *(complexio),* 22–23,
   31
Teodosi, Giovanni Battista, 39
Theophrastus, 47
therapy, 22, 23, 68, 69. *See also*
   bloodletting; *consilia;* purgation
   and purgatives
Tingus, Philippe, 11
Trincavella, Vettor, 10, 39
Trithemius, Johannes, 57
Turin (city), 73–74
Turin University, 64, 66–67, 77, 81

universities. *See names of specific
   universities*
Urbino, Duke of, 24
uroscopy, 43, 60

Varolio, Costanzo, 72
venesection. *See* bloodletting
Venice (city and Republic): epidemics in,
   21; government of, 15–16, 17, 20, 24;
   Lange's visits to, 41, 45; medical
   education in, 16, 100n3; and proximity
   to trade/travel routes, 19; publishing
   in, 26–27; religious repression in,
   35, 36
Vermigli, Pietro Martire, 19
Vesalius, Andreas, 82
Vettori, Piero, 25
Vienna, 16
virginity, 75–76
Vives, Juan Luis, 9

Waldkirch, Conrad, 28
Wechel, house of (Frankfurt), 11–12,
   84
Weinhart, Paul, 26
Wirth, Georg, 39, 45, 49, 52, 122n38,
   126n52
Wirth, Georg (elder), 40
Wirth, Michael, 40, 52
Wirth, Peter, 40, 43–44, 52
witchcraft. *See* magic (natural and
   demonic)

witness and testimony, medical, 8
Wittenberg University, 18
Wolf, Kaspar, 10

Zefiriele Bovio, Tommaso, 35
Zurich, 6, 20

Zurich University, 16
Zwinger, Theodore, 3–4; Augenio's
    correspondence with, 74, 145n46;
    Mercuriale's correspondence with, 17,
    27–28, *29*, 33, 34, 36, 74
Zwinglian theology, 6, 21